THE DISCOVERY OF
ANIMAL BEHAVIOUR

THE DISCOVERY OF

ANIMAL BEHAVIOUR

JOHN SPARKS

COLLINS
BRITISH BROADCASTING CORPORATION

Published by William Collins Sons & Co Ltd
London, Glasgow, Sydney, Auckland, Toronto, Johannesburg
and by the British Broadcasting Corporation
35 Marylebone High Street, London W1M 4AA

First published 1982
First reprint October 1982
© John Sparks 1982
ISBN 0 00 219061 3 (Collins)
ISBN 0 563 20090 1 (BBC)

Filmset by Jolly & Barber Ltd, Rugby
Colour reproduction by Adroit Photo Litho Ltd, Birmingham
Printed and bound in Great Britain
by William Collins Sons & Co Ltd, Glasgow

CONTENTS

INTRODUCTION

I write this in the Antarctic, not too far from where, in the shadow of the gently puffing volcano, Mount Erebus, the bleached huts once inhabited by Scott and Shackleton stand preserved by the frigid climate. Before me is an awesome, dazzling white vista of broken ice extending across the Ross Sea towards the mountains of Victoria Land. The air has such clarity that the eye is deceived in the estimation of distances. Those peaks over three thousand metres high appear a mere afternoon's walk away, but are, in fact, seventy-five kilometres to the west. In such a place as this, you might be forgiven for thinking that nothing stirs except the snow driven by the unrelenting polar wind, and the restless sea. And yet, here at Cape Bird, at the tip of Ross Island, just over a thousand kilometres from the South Pole, I can watch animals in action.

Two hundred metres away, a Weddell seal pup, which miraculously survives in its chilly cradle of sea ice, pesters its plump mother for food. She is dozing in the noon-day sun, but her baby snaps at her bewhiskered face and wails mournfully. She eventually relents, rolling languidly onto her flank to expose her nipples and offer a supply of warm, creamy milk. In a lead which has opened up in the ice, a lone minke whale surfaces for a second or two, lifts its flukes high in a flurry of cascading water, and sounds, leaving a cloud of steamy breath hanging in the air as the only evidence, tangible though ephemeral, of its brief appearance. Closer by, a more menacing animal lurks – a leopard seal whose beguiling smile belies its savage disposition. It has learned to patrol the edge of the ice where there is good traffic in penguins. At the northern end of Ross Island, there is no shortage of these quaint creatures. From where I sit, I can see the suburbs of a rookery of around 25,000 pairs of Adelie penguins. The odour has that suffocating mustiness characteristic of seabird cities; the air resonates to the clamour of braying and trumpeting birds vying for the choicest territories. Such a throng of penguins busily breeding is as impressive a sight as any naturalist like myself could wish for, and is the reason for my being here – in quest of film for the

Adelie penguin rookery on Ross Island in Antarctica

BBC television series, *The Discovery of Animal Behaviour*.

One of the first people to record this kind of bird was the French naval explorer, Durmont d'Urville, who in 1839 forced his way through the perilous pack ice until his frail wooden sailing ship, *Astrolabe*, reached the southern polar continent. In this uncharted area, he observed little black and white penguins waddling around on the ice floes like drunken sailors, and fondly named them after his wife, Adelie. He was not interested in their habits, but, like many early Antarctic explorers, saw them as a means of replenishing his ship's victuals. It was a century later before naturalists with scientific leanings visited the rookeries and, under more comfortable circumstances than the pioneering travellers experienced, explored in depth the Adelie penguin's customs.

As a maker of films about natural history, I use the research of such people, whose descriptions of the behaviour of these fascinating birds enables me to interpret what I see at a glance. By referring to a handful of scientific accounts, I can quickly 'read' an Adelie penguin's intentions, spot his battle cry, observe his mating display, and understand the significance of the raucous mutual ceremony which he and his 'wife' perform after a temporary separation. Engaged as I am in a form of zoological journalism, I rely greatly upon those whose unquenchable curiosity resulted in the charting of ever wider areas of knowledge. By edging towards the truth about why animals behave as they do, these naturalists and scientists provide the basic explanations for the moving images which the camera commits to celluloid.

And yet explanations, like fashions, change; what is accepted as satisfactory one day, may be cast onto the scrap heap of obsolete scientific ideas the next, in favour of a fresh version of the 'truth'. So it is with theories about animal behaviour. Take the Adelie penguins, for instance. The film about them will emphasise how the crowding of the nests foils the raiding skuas which take every opportunity to seize unguarded eggs and snatch unwary chicks. By nesting close together, the Adelies force the skuas to run the gauntlet of fiercely stabbing beaks, to the benefit of the penguin families. But a century ago, quite different explanations were in vogue. In those days, naturalists might have speculated about what was going on inside the penguin's mind, and have used behaviour to provide clues about the nature of the bird's 'thoughts and emotions'. A century before that, the Antarctic penguin would have been seen as a marvellous celebration of

God's wisdom in making an animal so well endowed with fat and feathers to withstand the rigours of the harsh climate. A medieval chronicler would doubtless have had his own version of the truth, spiced with myth and superstition about mysterious marine creatures displaying a peculiar mixture of 'fish and fowle'. Even in natural history, our vision of how things are is filtered through a glass tinted by the learning and prejudices of the age in which we live.

The Discovery of Animal Behaviour tells of the way we came to understand how animals conduct their lives. Although animals and their habits are unquestionably of central importance to the story, I have chosen to tell it through some of those people who, over the ages, came closest to understanding them. The narrative thrust is therefore carried along on the achievements of a selection of clerics, naturalists, psychologists, and zoologists who, by making penetrating observations or by conducting crucial experiments, advanced our knowledge on the subject. Like stepping stones, they take us along on a voyage of discovery, ranging from the fabulous natural 'mysteries' of the middle ages, to the present day, when electronic technology enables scientists to solve hitherto insoluble questions about the social life of animals. On the way, the story will uncover some of the mysteries of migration, reveal how the secret languages of animals were decoded, and show how some creatures can fool us into believing that they possess minds that work like our own. This 'natural history' will surprise professional zoologists and animal lovers alike by uncovering some of the strange circumstances which led to many major discoveries in animal behaviour.

The Discovery of Animal Behaviour was originally conceived as six one-hour films, each a blend of dramatic reconstruction and wildlife sequences aimed at a broad-based television audience. The flow of chapters in this book follows the progression of episodes in that series, rewritten to meet literary demands rather than the requirements of a spoken narrative.

It is often difficult when writing about animal behaviour to avoid ascribing human thoughts, motives, and feelings to our fellow creatures. To scientists, this is heresy. They take the view that, since we can never know for sure what animals feel and think – if indeed they do feel or think – then their behaviour is best described and interpreted in words that do not carry emotive nuances. Scientists have therefore coined their own terminology in an attempt to avoid the problem. And yet, their discussions are often beset with semantic difficulties as they wrestle with the problem of how to define the meanings of words in everyday use, such as 'intelligence', 'instinct', and 'territory'. Such arguments would be out of place in a book of this nature. So would scientific jargon. I have, therefore, chosen my language with the aim of achieving scientific accuracy, while engaging the interest of the general reader who is not over-concerned with academic objectives.

Like all television producers faced with making sense of subjects when rationed for air time, my problem was one of selection. The medium itself exercised its own brand of gentle pressure over who and what were chosen. Television thrives upon that which makes for compelling viewing, and is less satisfactory as a vehicle for conveying ideas which cannot easily be demonstrated visually. Those scientists who have made important theoretical advances in the subject, and who deserve places of prominence in definitive histories therefore, have tended to be omitted from the films. It also has to be admitted that some scientists have made great discoveries by experimenting on animals using surgical techniques. No amount of skilful faking for the camera would have generated sympathy for their work among viewers at large. In this book, I have been able to rectify some of the imbalances and omissions, and have been able to flesh out many details glossed over in the television series.

Although the warp and the weft of the story is mine, I have attempted not only to pick out the threads which interest me, but also to achieve a texture which reflects the pattern of scientific progress in understanding why animals behave as they do, and which conveys some of the excitement of discovering their nature.

John Sparks,
Cape Bird,
Ross Island,
Antarctica, 29 Nov. 1981

TO THE MEMORY OF
MY MOTHER AND FATHER
WHO ENCOURAGED MY INTEREST
IN NATURAL HISTORY

ANIMALS IN ACTION

In a jungle glade in Barro Colorado, Panama, a shaft of sunlight spotlights an intriguing natural drama. A large tropical spider stirs into action, and spins herself a web. With the greatest precision, she pays out a single filament of silk from her spinnerets, and fixes it to the spokes which she has constructed in a radial pattern from the hub. After nearly an hour of trailing her sticky thread round and round in ever-decreasing circles, her work is finished. The light glints off her delicate gossamer creation; but the net is as deadly as it is beautiful. She has slung her trap across a gap in the foliage through which insects regularly pass. Soon she is rewarded for her effort; a fly hits the web, and in a flash the hungry spider locates the position of the struggling insect, using the web's vibrations, which she feels through her feet. After biting the fly (which paralyses and ultimately kills it) and wrapping it in silk, she carries it to the middle of the web to digest her food at leisure. However, the flyway is busy, and within a second or two another insect has blundered into the net. This time, the female simply bites her victim and packages it, leaving it suspended in the edge of the net. She will return to it later after finishing her first meal. If she has a chance, for, unknown to her, another actor shares her silken stage. Unseen, a small spider enters from the wings. It treads carefully because the slightest clumsiness will set the trip lines shaking. The little spider is a thief, living by stealing insects from the larders of others. Its livelihood depends upon the utmost delicacy of touch. Its problem is to release the packaged fly by nibbling through the tough, tense strands holding it in place, while, at the same time, attaching its own tie-lines, which will spring the fly free of the plane of

The spider and the thief.
A small parasitic species lives in the web of a large Panamanian spider, and survives by stealing flies from the web of its giant host

the web when the last strand is severed. The intruder tiptoes around the web, using its long legs to maintain the vital tension while cutting each thread which anchors the trapped fly. Finally, the last one is parted, and the package swings clear. The thief can now consume its contents without fear of interruption.

The scene includes all the elements which make animals in action so enjoyable to watch. It is unscripted theatre. It has beauty and a touch of cleverness, and the stars exercise skills which leave us – the audience – open-jawed with amazement. The cast often performs a series of acts which combine into a satisfyingly dramatic shape. Predators and their prey masquerade before us, personifying good and evil; there is melodrama and tragedy in their eternal struggle. There is comedy in many courtship antics, pathos in parental behaviour. And there is often an element of mystery to engage our sense of wonder. But no matter how much pleasure we derive from their performances, animals are not in the entertainment business.

This begs the question: what is animal behaviour? Today we take it to embrace all those activities which enable creatures to survive. However, the term 'animal behaviour' is comparatively recent in origin, and was conceived within the framework of modern science. It appeared in the literature around 1900, when it was used by Conway Lloyd Morgan of University College, Bristol (now Bristol University). Before that, the word 'behaviour' was rarely applied to animals. Troops under fire behaved, as did iron filings in a magnetic field, and gas under varying conditions of temperature and pressure; animals, on the other hand, were thought to possess a psychology, and displayed 'habits, manners, customs, and instincts'. For the first time Morgan applied the term 'behaviour' to animals in a wide and comprehensive sense, covering the whole spectrum of their activities.

'Animal behaviour' defined a subject upon which the lens of scientific research could be focused. Gradually, some kind of order became imposed on a confusing array of animal habits. They became divided into categories, some corresponding to the necessities of life (e.g. feeding, grooming, breeding), others encompassing broader topics (e.g. navigation, communication, learning and instinct). This kind of ad hoc 'pigeon-holing' persists to this day, and is a useful means of organising the study of animal behaviour.

Locomotion usually plays an obvious part in the way an

animal adjusts to the world around it: a worm wriggles out of harm's way; a vulture soars in the sky, searching for tell-tale signs of a carcase; a zebra gallops in flight from a pack of hyenas. Animals employ many techniques for locomotion. For instance, jellyfish, scallops and squid propel themselves by 'jetting'; fish usually scull with their tails; turtles row themselves with their powerful flippers. On land, there are hoppers, walkers, crawlers and runners. A snail slithers along on its slimy 'foot'. Other creatures possess two legs, some four; insects employ six; millipedes, despite their name, move sedately on up to two hundred and forty!

Migration is one of the most impressive manifestations of animal movement. The world is criss-crossed with the highways and flyways of hard-travelling wildlife. In North America, huge flocks of Arctic-breeding waders and wildfowl follow regular routes across the continent to escape the cold blast of winter. Some of these avian journeys are astonishing. Each autumn, blackpoll warblers, which nest in pine woods on the edge of the tundra, assemble close to the New England coast to fatten for a daunting transoceanic flight. For a week or two, the birds devour as many insects as possible, doubling their weight to just over 23 grams (0.8 ounce). When a cool dry evening occurs, the featherweight warblers flutter into the air, climb slowly to an altitude of about 1000 metres (just over 3000 feet), and set out south-eastwards over the dark and forbidding Atlantic Ocean. On take-off, each blackpoll warbler carries a payload of fat sufficient to fuel its flight muscles non-stop for between four and five days. After that time it will have overflown Bermuda and the Antilles, and, with the assistance of favourable winds, made landfall 4000 kms (2500 miles) away in Brazil with its promise of bountiful supplies of food. The following spring, the warblers take a more westerly route over Central America back to the sub-Arctic pine forests. For a little bird which will fit comfortably in the palm of a hand the extent of its wanderings is almost miraculous. Arctic terns, though larger, are transglobal travellers on a scale unequalled by any other migrant. Some annually commute between the northern and southern pack ice, crossing the equator twice. The round trip must be at least 40,000 kms (25,000 miles).

A North American butterfly also migrates in a spectacular fashion, and provides one of nature's most unusual sights. Some monarch butterflies fly between two thousand and three thousand kms from the Great Lakes to over-winter in the

Autumn and spring migration
routes of the blackpoll warbler

Animals employ many
techniques for locomotion. One
of the most spectacular is flight,
as in the controlled wing-beats
and splayed feathers of this
little owl about to land

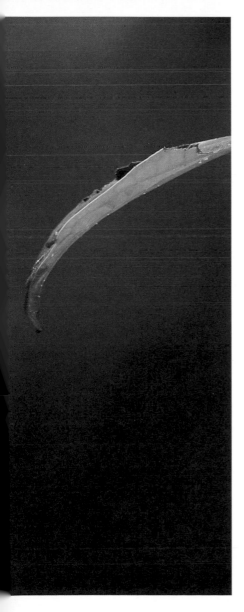

Another kind of movement. A tree frog, caught in mid-leap, shows the power of its thighs and legs

comparative warmth of California and the countryside around the Gulf of Mexico. Here they cluster on traditional 'butterfly trees', transforming the trunks into columns of golden wings, and festooning the branches with moving foliage. On sunny days, the trees come to life, as millions of monarchs sally forth across the glades in shimmering flurries.

Many mammals are migrants. In the far north of the American continent, caribou trek across the tundra between their calving grounds and the areas where they escape some of the rigours of the Arctic winter. Before the spread of European colonisation, dense herds of American bison undertook journeys across the prairies of several hundreds of kilometres, searching for fresh grass. However, the modesty of the distance covered was offset by the size of the herds. In 1871, one herd crossing the Arkansas River was a carpet of bison 80 kms (50 miles) long and 40 kms (25 miles) wide. Along the west coast of America, grey whales swim south from the plankton-rich Arctic Ocean to the tepid seas off California and Mexico to give birth to their calves. The round trip of 10,000 kms (6000 miles) takes approximately three months.

While on the move, animals often achieve remarkable feats of *navigation*. Some are more impressive than others. When covered by the rising tide, limpets become active, grazing on algae; but before they are left high and dry they accurately return to their home base. When exposed by the receding sea, periwinkles on the mudflats undertake round trips of a metre or two. Their trails reveal that the molluscs finish close to their starting point, thus maintaining their position in the inter-tidal zone. Although their eyes are minute, evidence points to the fact that they function well enough for the periwinkles to see landmarks, such as trees, cliffs, and even lighthouses, from which they take their bearings. Long-distance travellers, like birds, employ the sun and stars as well as prominent features of the countryside with which to set their courses. They also seem to possess a keen sense of relative time which enables them to compensate for the continuous movement of the heavenly bodies in the sky. To achieve this same end, human beings require chronometers! Recent research has revealed that a whole range of creatures, including bees, dolphins and many birds, are equipped with microscopic 'compasses' within their bodies, enabling them to sense the Earth's magnetic field. Despite the fact that the lines of magnetic force ripple and swerve over parts of the planet – as the capricious behaviour of

Millions of monarch butterflies cluster on trees at traditional over-wintering sites in Mexico. Some will have flown 3000 kms from Canada to settle here

Caribou migrating through the mountains of Alaska. They rest on ice to avoid troublesome flies

compasses demonstrates only too well – some migrating animals may use this sensitivity to orient themselves. Salmon are in addition able to 'remember' the taste of their home river, recorded in their 'memory' when as smolt they left for the sea.

Keeping clean and free of parasites is indisputably necessary for an animal's survival. *Grooming* is therefore a vital aspect of behaviour, appearing in different guises according to individual requirements. Most advanced animals comb, rub, scratch, or lick their coats. Herons, the danger of fish slime fouling their plumage ever present, possess powdery down which they wipe over their feathers to remove any such contamination from their bodies. Aquatic birds usually have well-developed preen glands, from which they smear oil over their feathers to help keep them water-resistant. Moisture loss is a problem faced by amphibians and insects: some spread waxy secretions over their bodies to cut to a bare minimum water loss through evaporation. Toilet behaviour is occasionally elaborate: a host of birds and mammals 'dust bathe', some birds even servicing their plumage with acid-squirting ants. Elephants plaster their massive bodies with mud; when it dries and falls off, it serves to remove unwanted ticks. Grooming can be a social event: many birds and primates spend a great deal of time picking through each other's fur or feathers. Reef fish even possess their own 'barbers' – small and brightly-coloured cleaner wrasses which station themselves at special places to attend to the skins of their clients. So important is the task of keeping clean that even apparently helpless animals have evolved remarkable methods of keeping unwelcome parasites at bay. Between their spines, sea urchins possess a formidable array of tiny claws, which crush, cut, or inject corrosive chemicals into anything that settles on their surface.

The eternal quest for food provides fascinating examples of animal behaviour. In fact, almost every imaginable method of *feeding* is employed by some animal somewhere. The larval ant lion digs a pit in which it ambushes small insects scurrying across the sand. Praying mantises sit and wait until an unwary fly lands near them. Some creatures use lures; the wriggling tongue of the alligator snapper bears an uncanny resemblance to a juicy worm, and tempts small fish into the turtle's jaws. Many herons encourage fish to come within striking distance of their beaks by spreading their wings like a parasol, providing their prey with relief from the heat of the sun. For a mollusc, the cuttlefish is an ingenious killer. Cruising along the sea bottom, it directs a jet of water onto the sand to uncover hiding shrimps. When one is revealed, it can make no escape. A pair of long suctorial tentacles envelop it in a flash. Among birds the variety of the structure of their beaks reflects the diversity of their feeding habits. The massive bill of the hawfinch can crack cherry stones; the flamingo's bill is a refined sieve, adapted for filtering minute shrimps from the surrounding brine; humming-birds live by sucking sugar-rich nectar from flowers and in consequence their beaks are built rather like drinking straws. Birds of prey possess meat hooks; woodpeckers, chisels. Some species, such as ox-peckers, seek their food on the surfaces of other creatures. Bats use sound to echolocate flying insects – and some moths have even evolved devices for jamming the bats' 'radar'. Dolphins bounce sounds off underwater targets, and can tell from the nature of the echo whether the object is solid and stationary like a coral head, or whether it is composed of flesh and bone and moving like a fish. A dolphin can even stun its prey with a very intense burst of sound. Certain fish use pulses of electricity, not only for navigation and communication, but also for locating food. The electric eel and the torpedo ray, both highly charged, employ bolts of electricity to kill, or at least to immobilise, their quarry.

A few creatures even use tools to enhance their feeding efficiency; song thrushes shatter the shells of snails by hammering them on stone anvils; sea otters often carry their own personal stone anvils around with them, which they use to break open shellfish; the Egyptian vulture smashes ostrich eggs with the assistance of well-aimed stones, and chimpanzees 'fish' for termites with twigs. Other animals store surplus food supplies against a future need: squirrels hoard nuts; foxes bury excess prey; pikas, small rabbit-like rodents inhabiting mountainous regions, build haystacks during the late summer to dry their winter fodder.

Chimpanzee using a twig to 'fish' for termites. Perhaps the first one to do so chanced upon this behaviour by accident

Overleaf: Trapped by a tongue. A chameleon demonstrates its extraordinary feeding technique – its prehensile and sticky tongue can uncoil to a distance equivalent to half its body length

Many varieties of animals must protect themselves as best they can against predators. They possess ranges of behaviour exclusively adapted for *defence.* Occasionally, evasion involves the total cessation of movement, as in the case of the superbly camouflaged eider duck which crouches motionless over her eggs when she senses danger. Defence tactics are occasionally dramatic; the bombardier beetle, for example, generates in a

strengthened chamber in its tail an explosive mixture of chemicals, which can be directed at its attacker. Guard bees and wasps display suicidal tactics, acting like kamikaze aircraft if their nests are raided. Soldier termites of certain species have heads shaped like cannon; when confronted with intruding ants, the termites are able to project gum from their 'snouts' to ensnare the marauders. A small East African termite called *Trinervitermes* has no special caste of soldiers. For defence against ants, the workers have the capacity to commit a kind of hara-kiri; their bodies rip open along a pre-formed line of weakness, releasing a sticky resinous substance which the ants find unpleasant. Some otherwise cryptic moths and mantises display frightening 'eye' spots on their wings when disturbed, to deter a predator from pressing home an attack. Mass mobbing is also a powerful method of disconcerting an enemy: Arctic terns, for instance, are able to draw blood even from human intruders. And nothing matches the roaring charge of a fully-grown male gorilla, which is designed to terrorise the most stout-hearted of opponents. Although the charge is mostly display, a desperate gorilla is capable of pressing home an attack, to wreak frightful damage on an enemy.

Breeding generates a great deal of dramatic and eye-catching behaviour. The males of many species become landowners, and the competition for mates and space often goads them into gladiatorial contests. During rut, bighorn sheep batter their heads together with such force that the brain-numbing reports echo across the North American mountain valleys like rifle shots; bull elephant seals hack and slash at their opponents' necks until the flesh is raw and drips with blood; rival cock robins thrust and parry with their beaks as they stake their territorial claims.

Breeding also involves the performance of a bewildering variety of *communication* rituals, often designed to placate the female's fears and soothe the way to mating. Fireflies flash signals, each species with its own code of long and short bursts of light; some spiders use semaphore techniques, while others woo their mates by drumming a precisely orchestrated tattoo on their webs — wolf spiders appeal to the appetites of their ever-hungry mates by gingerly offering them food parcels: gift-wrapped flies! For sheer extravagance, nothing exceeds the cock birds of paradise, which parade in unmatched splendour before their dowdy hens. Many species broadcast their

Grooming is a vital aspect of behaviour. A harvest mouse licks its tail, a cleaner wrasse attends to a coral trout, while two rockhopper penguins indulge in mutual preening

A pair of tropical frogs mating

Overleaf: The race to mate. Bighorn rams batter their heads together – the winner will attain priority in the mating stakes

desire for sex through sound alone: crickets scratch out urgent trills by rubbing their wing cases together; cicadas produce an insistent rhythm on miniature drums set into their bellies; some fish squeak and grunt with the assistance of their swim-bladders; higher vertebrates use their voices to make music. Chorusing tropical frogs can create a deafening cacophony, but a few birds, such as the European blackbird and nightingale, can produce songs of exquisite beauty. By contrast, most mammals are less appreciative of tunes (although gibbons and whales are as melodious as the best of avian songsters) and tend to communicate passion and territorial ownership by means of scents of varying degrees of pungency. However, no matter what methods animals use for courtship, every variety of such behaviour leads to the all-important assignation between the sex cells.

A multitude of *mating* conventions exists to ensure that eggs and sperm meet, thereby guaranteeing further generations of living creatures. Some of the mating techniques found in the world of nature make human lovemaking look quite conservative! Many animals without backbones indulge in solitary sex, pouring out milt and spawn directly into the sea. As a consequence, sea urchins, starfish, most aquatic molluscs and primitive crustaceans lack companionship during the most important event of their lives. Some marine worms, of which the Pacific palolo worm is the best-known example, are exponents of 'packaged sex': they confine their eggs and sperm to the rear parts of their bodies, which are then detached and shaken free to swim around in the plankton. When male and female portions happen to meet, the body-walls rupture, releasing the sex cells into each other's company. Octopuses and squid also wrap up their sperm in rather complicated explosive packages; the male delivers them to his partner by means of a unique 'mating arm'. In the case of the paper argonaut, the males are diminutive and possess a detachable mating organ which, after grasping a bunch of sperm packets, wriggles away, and homes in on willing females by itself! Among vertebrate animals, even amorous male newts practise packaged sex: each courting male lures his female over the spot where he has deposited a semen-filled envelope by wafting an irresistible chemical towards her with his tail. As she passes over the 'love letter', her mobile ventral lips embrace the package and draw it into her genital tract. Most terrestrial creatures have evolved intimate techniques of sexual engage-

Undersea battle

A red anemone about to attack its neighbour

It lunges forward

And makes contact with its stinging tentacles

The victim responds as the aggressor presses home the attack

ment which enable the process of fertilisation to take place safely, deep inside the female's body. Even so, copulation comes in various forms, from the brief genital contact of birds to the prolonged mating lock practised by members of the dog tribe. Usually consenting couples attempt to mate in some kind of seclusion, but some specialise in sexual orgies – many ponds in spring become filled with great writhing balls of spawning toads.

Certain animals are equipped with both male and female sexual organs. Earthworms, slugs, and snails which can often be observed mating in gardens are hermaphrodites. The colourful wrasse, *Anthias squamipennis*, which swims above the coral gardens of the Red Sea, even changes sex as it grows. Behaviour is the key to the sexual transformation. Every *Anthias* is born a female, and joins a shoal of golden females dominated by a single male. As she grows, she advances in rank because she can bully those smaller than herself. When she is at the top of the pecking order, a remarkable thing may happen. Should the male disappear, she may smartly take his position, and, over the course of a few days, become transformed into a male.

Parental behaviour embraces all the manifold techniques employed by animals to make provision for their offspring. This may entail nothing more than the mother choosing the correct food plant on which to lay her eggs: for example, peacock and monarch butterflies deposit their clutches on stinging nettles and milkweed respectively. Some kind of care is usually lavished on the eggs or brood. These may be guarded (python), carried around – on the back (some toads), beneath the belly (prawns), in the mouth (certain cichlid fish) – or actually incubated: crocodiles, for instance, use the heat of the sun. Brush turkeys dig pits to employ the heat of rotting vegetation or even the warmth of volcanically heated soil; the majority of birds, however, sit on their eggs to keep them at a viable temperature. Food is often brought to the brood. Some young spiders are nourished by the corpse of their mother who dies after laying her eggs. The fry of the South American *Discus* fish also feed off their parents, but in a less drastic manner – the parents produce a highly nutritious slime from their skin, on which the young graze. Birds whose nestlings are helpless bring back food to them, sometimes partly digested in their crops. Sandgrouse, which breed in areas baked by the sun, even transport water to their chicks in their breast feathers, which they soak

Slugs mating in mid-air, suspended on a thread of mucus. Like most molluscs they are hermaphrodite, but need to exchange sperm in order to breed

while taking their daily drink. Pigeons manufacture 'milk' in their crops for their very young squabs; mammal mothers make nourishing fluid, true milk, in special glands slung from their torsos. In making nests in which to harbour their young, animals prove to be talented architects. Siamese fighting fish build rafts of bubbles for their eggs; termites erect huge air-conditioned tenements in which to house their fungus gardens and brood chambers. Weaver ants use their larvae like 'needles', employing the silk secreted by the grubs to 'stitch' leaves together.

As young creatures grow up, their behaviour develops, through *'instinct'* and *'learning'*. Both of these processes, together with all matters relating to *'intelligence'*, fall within the province of animal behaviour.

Instinct reveals itself in the way that some behaviour patterns are performed perfectly from the first. The complicated rituals of courting birds are cases in point. The young of hoofed animals are able to gallop faultlessly alongside their mothers, shortly after birth. Mason wasps fashion their nests out of clay without benefit of instruction. Young cuckoos navigate from Europe to Africa by themselves, their parents having migrated several weeks beforehand.

Despite the advantages of inborn skills in preparing animals for survival, all creatures need a touch of 'virtuosity', or learning, in order to cope with the world which is ever presenting them with surprises, challenges, and opportunities. For example, the relentless quest for food has led British titmice into acquiring all sorts of tricks. They have developed a flair for following milkmen and perforating the caps of milk bottles on doorsteps in order to drink the cream. Since 1928 when the annoying behaviour was first recorded in Southampton, the tits appear to have learned from each other, and now the habit has spread throughout the kingdom and over much of Europe, to the benefit of the birds.

Cleverness of this kind is common in the living world, and has led people over the ages to become obsessed with the nature of animal intelligence, and with the need to discover what goes on inside the animal *mind*. Attempts at constructing scales of intelligence have generally floundered because of the impossibility of devising tests which can compare, in any meaningful way, the mental prowess of different species. Although animal lovers still debate which species ranks second to man as the most 'intelligent', the most sensible view is that

Most animals lavish some kind of care on their offspring. Here, a mother cat carries her kitten to safety

animals are generally equipped by natural selection with sufficient brain power to allow them to survive in their own particular niches. Living on a diet of fresh *Cecropia* leaves, the three-toed sloth does not need a dazzling intellect; dolphins, on the other hand, hunting fish in packs, must be quick-witted and socially aware.

It is little wonder that many modern zoologists have tended to avoid such a contentious subject, and, instead, have opted for more tangible areas of research, such as animal *societies*. Within the animal kingdom, a great many types of communities can be found, from the loose aggregations of feeding winkles clustering on kelp stranded on the seashore, to the highly-structured societies of ants, bees, and termites with their well-differentiated caste systems. Some otherwise solitary species, such as spider crabs, come together only to breed, while troops of monkeys and apes stay together year after year. Some social gatherings are composed of both sexes; others are single-sexed. The way in which wild creatures organise themselves into herds, troops, flocks and shoals is the outcome of behavioural processes, and so within the legitimate sphere of animal behaviour.

Such knowledge as we have about animal behaviour has been discovered and described by naturalists and scientists who have a professional interest in taking some of the 'mystery' out of it. They have addressed themselves to revealing what causes animals to behave as they do, and to observing how they actually act. Some of these nature detectives assume the role of hunter; they observe animals in their wild haunts, spending long periods squinting through binoculars, trailing over bogs, climbing trees, crawling through tangled undergrowth, ascending in hot-air balloons, and scuba-diving into the cool depths of the sea in order to get good views of their quarry. Other investigators are content to work in the relative convenience of laboratories equipped with scientific instruments which automatically spew out huge amounts of data, peering at creatures as they scamper through mazes and press levers with the skill of well-trained circus stars. Both kinds of exploration fall within the framework of the scientific endeavour which has revealed the secrets of animals in action. The naturalist with a love for the open air and the scientist who beavers away beneath the glow of fluorescent lights both attempt to understand how animals live and learn.

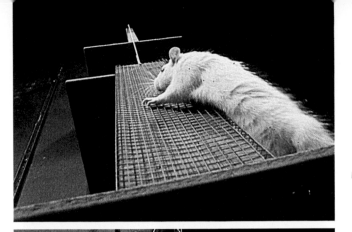

Virtuoso performance

A rat demonstrates what it is capable of through training. It begins by climbing the ladder

Having reached the top of the ladder, it is now on the middle platform

The rat then hauls up the ladder in order to climb to the next level

It can now reach the food on the top platform

Every new fact helps to sharpen our perception about how behaviour assists animals to survive and to thrive. Hardly a day goes by without new information being contributed by an army of researchers who are now employing all kinds of techniques to make their discoveries, as a few examples will show.

Dr Sandro Lovari, a zoologist from the University of Siena, is fascinated by the behaviour of chamois, and uses the traditional methods of a field naturalist to record the details of their private lives. Armed only with note-books and pencils, and dressed in combat camouflage, he regularly makes the gruelling climb into the peaks of the Abruzzo National Park, on the roof of Italy, where his quarry live. Early in November it is a desolate place, with flurries of snow sweeping across the amphitheatre of rock. This is when the chamois stage their annual rut, the males chasing each other up and down breathtaking cliffs in order to attain priority in the mating stakes. In spring, the scene changes dramatically, and the alpine meadows are dotted with gentians. The warm sunshine drives sex from the minds of the chamois. Dr Lovari can then analyse the behaviour of the kids conceived during the previous year, as they play king-of-the-castle on outcrops of stone, and join their parents gambolling down drifts of the late-melting snow lingering in the mountain shadows. Patience, persistence, and a sympathetic understanding of his animals enable Dr Lovari – and many like him – to get results with a minimum of equipment.

Even insects can be studied in similar fashion. Dr Geoffrey Parker of Liverpool University spends a great deal of the summer gazing intently into cow pats because these are chosen by dung flies as arenas for their breeding rituals. Long hours of patient observation have revealed what is happening in the frenzy of flies which settle on each fresh heap of aromatic dung. Although it is the cow's waste, the warm slurry is highly nutritious, and the gravid female flies home in on the fresh pats in order to lay their eggs. This is also where the males lurk to catch a mate. But each male is faced with a dilemma: if he remains coupled to one female for too long, he will lose the chance of embracing others as they arrive eager for sex; if he mates too rapidly, he may be cuckolded by other males whose sperm may reach the eggs first. Dr Parker calculated that the optimum mating time for a male dung fly, in order to maximise his chances of fathering as many of a female's offspring as possible, should be forty-one minutes. That agreed reasonably

Mating time is crucial to a male dung fly. If too short, he fails to fertilise the eggs. If too long, he loses chances with other females

well with the average duration of copulation actually observed in the meadows around Cambridge – thirty-six minutes.

Many animals are able to convey messages to each other and to signal their intentions by means of body language. For instance, gulls on their nesting grounds face each other, sometimes with their heads held low, necks canted almost to the horizontal, and on other occasions with necks stretched tautly vertical. Gull watchers speculate that the postures communicate whether the birds are fearful and likely to flee from encounters or are brimming with self-confidence. Such theories have been put to the test in a rather unusual way by Professor John Stout. The glaucous-winged gulls, which nest on small islands off the coast of Washington State in the USA, regularly have to contend with a remotely-controlled robot gull which trespasses into their territories and challenges them with provocative postures. At the beginning of the breeding season, the birds are emotionally highly charged, and take the robot seriously, as though it were a genuine gull with blood and not electricity coursing through its veins. The set of its head and neck can be altered by a flick of a switch in the control panel. By looking at the effect on the gulls of the robot's poses Professor Stout can tell which behaviours are aggressive and which displays are taken as appeasement gestures, thus confirming the results gleaned from watching the behaviour of real gulls reacting to each other.

Electronic models have also been put to good use in Iceland. Harlequin drakes are lavishly coloured birds whose blue and orange plumage is dissected with black and white patterns. During the breeding season, they space themselves out along the banks of fast-flowing rivers like sentinels, and make themselves conspicuous with curious head-bobbing movements. A team of British scientists led by Dr Ian Inglis thought that this behaviour repelled other drakes, and caused them to keep apart. To test the idea, the team built a remarkably convincing mechanical drake which faithfully reproduced various kinds of head movements to order. When located on a river favoured by harlequin duck, the Laxa close to Lake Myvatn, the resident drakes appeared to be fooled by the dummy; when it bobbed, they rocked their heads in response, and kept their distance from the established intruder.

Trying to decipher the vocal language of birds can lead people to use unusually bizarre methods. Someone watching Lisalotte Hagleitner deeply engrossed in her research could be forgiven for thinking her a little crazy, because towards the

Body language – two typical herring gull postures. The 'upright' threat posture (above) communicates potential aggression, while the 'grass-pulling' behaviour (below) may also do so

end of the month of May, she talks to goose eggs! But there is method in her madness. She is interested in the calls of geese. Ganders start a dialogue with their goslings while they are still inside the eggs, so that by the time they emerge from the shells, they recognise their parent's voice. Lisalotte Hagleitner is a substitute mother, and encourages the unhatched goslings to pipe into her microphone by speaking softly to them.

Technological progress enables contemporary zoologists to discover details of animal habits denied to former generations of scientists. Bear behaviour is a case in point. Despite their bulk, bears are shy and mostly solitary beasts, and roam around in woods inconveniently dark and remote. Until quite recently, their movements and manners were almost impossible to monitor. Nowadays the task is incomparably easier than before because bears can be bugged with small radio transmitters concealed in collars. Dr Lynn Rogers of the US Forest Service has logged more than one thousand black bear captures, and has managed to collar every bear in his one hundred square mile study area in northern Minnesota. A crucial part of his work is the yearly visit to every bear's winter den to renew the transmitter batteries. In March when the snow lies thick and crisp on the ground, he tracks down the hibernating bears by following the telemetric 'pings' on his receiver. When he reaches each den, he hauls the slumbering bear out, weighs it, and its cubs if there are any, changes the batteries and takes a blood sample or two for good measure. Some people might take the view that this kind of research is an unwarranted invasion of the bear's privacy. But Dr Rogers has been able to find out many new facts about bear families. She-bears share their home ranges with their daughters, who then tend to inherit their mother's 'property' when she dies. Young males, on the other hand, take off into the forest and seek their own territories wherever they can squeeze in. Their land requirements are much larger than those of she-bears. However, females tend to settle on beats which overlap the territories of several males, presumably to have a choice of mates. Black bears prove to be very adaptable mammals, and the way they organise themselves tends to depend upon how the food is distributed. Normally they have to spread themselves out, but the presence of a good productive rubbish dump tends to concentrate bears wonderfully!

Bird tables laden with household scraps also bring hungry animals together, often in fierce scrummages. These are exaggerated versions of situations which many birds face daily in their search for food. Among finches, tits or chickadees, which habitually make up search parties of mixed species, the smallest birds may have to defer to their larger and more forceful companions. Some of the more easily dominated species have therefore evolved their own tactics in order to procure a reasonable share of the available food. The dainty marsh tit is always low in the pecking order. Its strategy for survival in a feeding fray is to nip in quickly, snatch a morsel of food, fly off, store it, and go back for another – and so on. Later it may return to its caches to feed at leisure, without having to fend off its larger companions. Nuclear technology is helping Oxford zoologist, Dr John Krebs, to discover the effectiveness of the marsh tit's furtive feeding habits, and to assess the accuracy of its memory. In a wood just outside Oxford, he dispenses harmlessly radio-active sunflower seeds, and, with the assistance of a geiger counter, finds where the birds hide them. Surprisingly, a marsh tit is able to conceal several hundred seeds a day, each one in a different place – beneath a loose piece of bark, under a leaf, or slotted into a cushion of moss. Furthermore, the nuclear-assisted nature detective has discovered that the marsh tits possess amazing powers of recall of where they have lodged their food. From his observations, it is clear that the tits eventually rediscovered virtually all of the seeds which they secreted away.

People like Dr Krebs are products of twentieth-century science, and it is their discoveries which have largely fashioned our contemporary attitude towards animal behaviour. And yet, what we know about animals and their habits was not always governed by the school of scientists who seek to base their findings upon demonstrable facts and figures. Those who today attempt to make sense of animal habits inherit a long and distinguished tradition, stretching back over thousands of years, accumulated by all sorts of different kinds of people who made the customs of living creatures their business.

Before the dawn of history, hunters depended on the accurate prediction of the habits of their quarry – as they still do. Farmers, fishermen and gamekeepers have been compelled to develop working relationships with their animals in order to manage their stocks successfully. Even the common people have observed the actions of the animals around them; but their livelihood did not depend upon accurate interpretations of what they saw, so they fostered many odd notions about animal mentality, a few of which linger even today.

But it was scholars, often monks, who saddled themselves

with responsibility for recording 'all knowledge', and this included what was known about the wild world. The story of animal behaviour largely embraces their work and its consequences, because it opened up the way both to modern natural history and to biological science, two disciplines which have gone far to illuminate the minds and manners of animals.

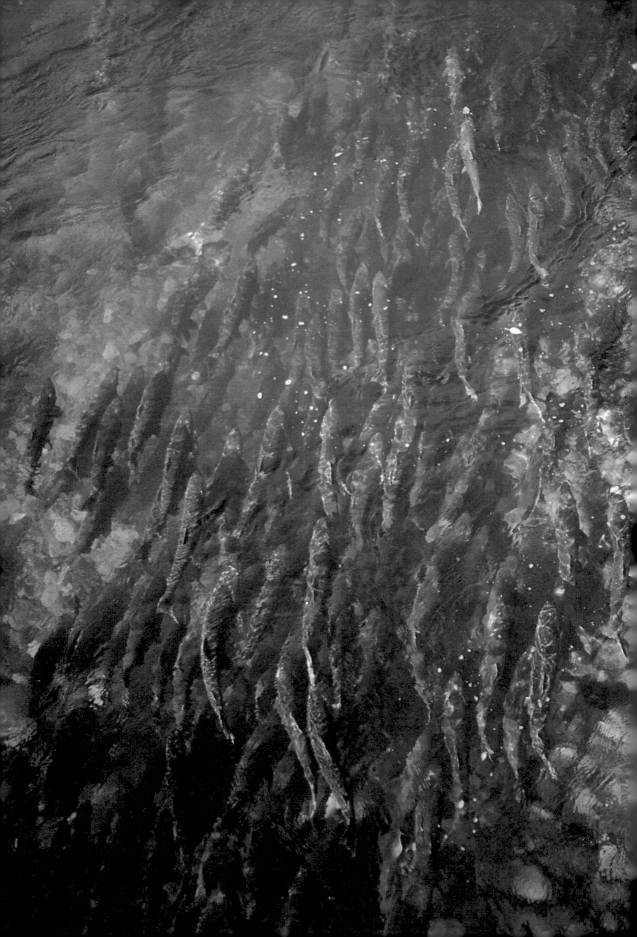

FACTS AND FANTASIES

Two mythical creatures: a sea monkey and a sea Turk. From a medieval woodcut

Some creatures perform extraordinary feats that beggar description and baffle our imagination. It is therefore little wonder that in their ignorance our ancestors often conjured up far-fetched explanations for what they saw, or thought they saw. Even nowadays, when we are comparatively well informed about the world about us, the common, everyday interpretations of animal behaviour reveal very little scientific insight.

Pets apparently 'know' what their owners are thinking; dogs which open doors are referred to as being immensely 'intelligent'; a purring cat is a 'happy' one; a talking parrot is a 'genius'.

And yet there have always been people such as farmers and hunters whose professions have brought them into close contact with both wild and domesticated creatures. Such people developed a good practical knowledge of behaviour, at least of those species with which they worked. But their interest in animal habits was purely empiric; what they learned was applied to making them better farmers, or more skilled at hunting. They were not concerned with the relevance of behaviour to the survival of the animals themselves.

The phenomenon of migration provides a good example. Today it is understood as part of an animal's strategy for survival. Though for some species the effort of travelling is tremendous, on balance the benefits outweigh the risks of sedentariness, and the consequent danger of starvation. However, in the past, people whose livelihood depended upon the seasonal appearance of wildfowl, fish, and big game were unconcerned with scientific explanations for the mystery. The

A shoal of migrating sockeye salmon. The livelihood of many North American Indians once depended upon the yearly occurrence of this vast abundance of flesh and oil

stone-age hunters of Europe plundered the processionary herds of bison and mammoth, and caught migrating salmon in the rivers. Unfortunately, they left us only their beautiful cave art as a cryptic record of their beliefs. Although we can never know for sure what these people made of the overnight appearance of certain animals in their countryside, and of their later abrupt disappearance, some contemporary tribal hunters can provide us with some clues.

A Haeda Indian salmon motif

The natives of the north-west Pacific coast of America call themselves 'The Salmon People'. For the past nine thousand years, until quite recently, their livelihood depended upon a yearly glut of food as various kinds of salmon ascended the rivers to spawn. Among them was the sockeye; the adult sockeyes spent several years feeding in the Gulf of Alaska, then swam to the gravel beds where their parents had bred several years previously. Resplendent in their vermilion reproductive colours, the spawning salmon clogged the rivers and, when they were spent, they died. The Indians were perfectly well aware of these basic facts of fish biology, and yet they embellished them with superstition, and imbued the salmon with spiritual qualities.

Like most people living in pre-scientific cultures, the Indians saw the world as a unified whole in which everything, themselves included, was involved in the drama of life. They were in no doubt that the fish surged into the rivers as an act of charity by the great salmon spirit, enabling the tribal people to benefit from the deluge of oil-rich flesh. Like many hunters, the fishermen had great reverence for the natural resources which were vital to their culture. Living creatures therefore commanded a degree of respect worthy of fellow inhabitants of the countryside; many, like the salmon, were endowed with conscious spirits. However, those of the salmon were known to be fickle. The Indians were never quite sure of the future because the numbers of salmon varied widely from year to year: sometimes the migration yielded fish in unimaginable abundance; on other occasions the catch failed, with disastrous consequences for the Indian economy.

Each year, the arrival of the first salmon was greeted with joyful celebrations by the Indians. Rituals were performed, designed to please the salmon spirits, and to ensure a repetition of the run the following year. If the correct customs were observed, the fishermen hoped to influence the size of the influx of salmon. Bones were returned to the river to ensure that the fish would be reborn.

Something of the rapport which the expert Indian fisherman had with his prey is reflected in the following prayer:

'Welcome swimmer. I thank you because I am still alive at this season when you came back to our good place; for the reason why you came is that we may play together with my fishing tackle. Now you go home and tell your friends that you had good luck on account of your coming here, and that they shall come with their wealth, that I may get some of your wealth.'

People, like the Kitsumkalum from British Columbia, Canada, therefore interpreted the salmon migration as the annual fulfilment of a contract between the hunter and his prey. Seen from this point of view, salmon behaviour was designed largely for the benefit of the Indians, rather than for the fish themselves.

The native Americans, like most people with their roots firmly embedded in the countryside, fully appreciated the interdependence of everything in nature. They thought of themselves as part of the living jigsaw, and sought inspiration in the things they saw and experienced to help explain the world. For example, the behaviour of the beaver suggested to the Indians a version of the creation. Beavers abounded in the waterways and always seemed to be busy constructing dams and clear-felling stands of cottonwood trees, altering the landscape in a most impressive way for so small a creature. This led some Indians to compose a charming story in which the world was fashioned by the industrious efforts of a mighty beaver. Although this particular creation myth is widespread in Indian lore, probably no one took it as literal truth. It was merely a useful metaphor based upon accurate observations of one of nature's architects, enabling the people to answer questions relating to the origin of the world.

Those who live in cultures that are close to nature tend to be preoccupied with the similarities between humans and animals. In the Pacific North-West, members of the Nootka tribe identified strongly with the timber wolf. The Nootka were hunters and had to survive on land over which wolves roamed. They therefore had to share many similar problems in tracking down game, and in providing for their families when conditions were grim. The Nootka had enormous respect for the wolf's prowess as a killer, and envied his stamina, stoicism, and uncanny knack of padding silently through the forests. Like the tribal hunters, the wolf drew much courage from the pack, and displayed a keen sense of territory, from which strangers were excluded. The Nootka also appreciated the wolf's fierce loyalty to the family. They noticed that wolves indirectly provided food for the whole community. After the pack had dined, the nutrients from the carcase seeped into the soil and stimulated a nourishing growth of grass upon which deer fed. Morsels left by the wolves were eagerly taken by scavenging foxes and crows. The hunting people admired, and attempted to emulate, all of these qualities. Some took the wolf as their totem, so establishing a kinship between man and beast based firmly upon behaviour.

In Europe, hunting has not been crucial to man's existence for thousands of years, and consequently we have lost the intimate relationship with wild creatures that our Neolithic forebears probably possessed. Nevertheless, when hunting equipment was laid aside in favour of the harness and the plough, farmers in particular needed to know how to manage their domestic animals. For this, they needed to understand the habits of their charges.

Controlling flocks of sheep is a tiring business, so shepherds enlisted the help of domesticated wolves – dogs – to do much of the leg work for them. They exploited the wolf's natural hunting behaviour. Often working in pairs, their dogs worked as though the sheep were deer or caribou. They stalked and circled them, crouched if the sheep looked too alarmed, and, if necessary, jumped at their throats in order to turn the 'quarry'. The dogs' behaviour remains the same to this day. Of course the shepherd cues his dogs by whistled or spoken commands. But in one important respect, even well-trained sheepdogs will rarely go against their ancestral hunting instincts. Like wolves, they need a pack leader and they treat the shepherd as such, always tending to drive their 'prey' towards him. Only with the greatest of difficulty can sheepdogs be instructed to take a flock of sheep away from the shepherd. It is simply against their wolf-like nature.

Where wolves or even feral dogs were a danger to sheep, the shepherds needed guards. In southern and central Europe, Anatolian sheepdogs have been selectively bred for the purpose. They are huge shaggy animals which, at a glance, even look like sheep. Indeed, the shepherds say that their dogs think they *are* sheep! This is not surprising since the puppies are brought up with lambs, and are not fondled by their human masters. When grown up, they mix with the flocks and treat the sheep as pack companions, even to the extent of romping with them and attempting to mate with ewes. Should

A Hungarian Kommondor, an aggressive guard dog that defends sheep by mingling with the flocks

Birds mob owls, a habit exploited by medieval bird catchers to lure small birds onto sticky perches. The technique is still used in Italy

a stranger, or another unknown dog, appear on the scene, the guards materialise from the mass of sheep in order to defend their 'pack'.

The control of domesticated farm animals has always called for a basic understanding of their behaviour. Throughout history farmers have ignored at their peril the bad temper of bulls; an early lesson in cattle husbandry must have been not to place two bulls in the same field if there were cows in heat anywhere about! Farmers must have watched their quarrelsome cockerels round up their harems of hens, and, at roosting time, flutter up to the rafters as though seeking out safe perches in trees as their ancestors did in Asian jungles. Without troubling with the origin of this behaviour, those who kept poultry furnished their birds with high roosting perches. This kept the birds happy while foiling foxes and rats.

Medieval bird catchers employed various tricks which indicate that they must have known the habits of their quarry. For example, some plied their craft with the help of an owl or two, relying upon the fact that many birds are infuriated by the sight of one of these nocturnal predators: song birds will throw caution to the winds, and will mob an owl in the open mercilessly, while hawks display their dislike by stooping on them. Eagle owls were favourite decoys. Their extravagant ear tufts and flaming orange eyes were especially effective in causing birds to explode with rage. One method used by the bird catchers was simply to tether one of these 'super-owls' to a block and surround it either with twigs covered with bird lime, or with a high net draped around the branches of nearby trees. The decoy was often placed on a perch projecting from a hide in which sat several men, each brandishing long cleft sticks for gripping the feet of unwary birds. During the fifteenth century, an amusing portable hide was used in Germany. It had a man inside who walked it around the hedgerows looking like a miniature moving hay-stack; on the outside was an eagle owl and a cleft stick which was operated by the incarcerated fowler as soon as a luckless bird happened to alight on it to scold the decoy. Today owls are still used in this way in countries such as Italy, where each year an estimated 150 million song birds are lured to their doom.

Modern sportsmen still keep some of the hunters' traditions alive. Wildfowlers must pit their wits against wary and fast-flying birds. They are able to deceive sharp-eyed ducks and geese with quite crude models, tempting them to land within gun range. The fatal response depends upon the tendency of

The behaviour of wild boars is accurately portrayed, showing suckling, feeding and copulation, in this painting from a medieval manuscript. Below, a stag hunt. A practical understanding of animal behaviour enabled hunters to pursue their quarry successfully

A fire-breathing dragon. From
an illuminated manuscript

birds to hanker for each other's company: the sight of a resting flock of ducks or geese bobbing up and down on water is irresistible. By trial and error, anglers too have found that fish can be tricked into attacking a very rough approximation of their real prey. A polished metal spinner glints in the water like a small fish, and a delicately twisted feather can be made to rest on the surface of a chalk stream as though it were a mayfly. When married to a barbed hook both are deadly, to pike and trout respectively.

By rights, the science of animal behaviour should have originated with hunters and those concerned with animal husbandry. It did not do so because farmers and their sporting brethren failed to record their accumulated wisdom. The influence, and interest, of scholarship was required. This was obtained when an intimate knowledge of animal habits, already an essential aid to survival, became also an intellectual pursuit. In this process, medieval farmers played a crucial role. As they became more skilled, they produced ever-greater surpluses of food, thus freeing many people to pursue knowledge for its own sake. This had been the case nearly 2400 years ago in Greece, where lived one of the greatest of all naturalists.

Aristotle can claim to be the first person to watch animals objectively. With almost nothing in the way of reference work to consult, he set about observing and recording what was going on in the world. Unlike farmers and hunters, whose perceptions were subordinated to the purely practical considerations of their work, Aristotle searched for a 'scheme of nature', and busied himself arranging knowledge into systems of classification. In his encyclopedia of zoology, *Historia Animalium* – best translated as 'Inquiries into animals' – he employed a 'ladder of life' in an attempt to illustrate the relationships between different living creatures: at the bottom, he placed inanimate matter and lower plants, and at the top, man. His ladder anticipated the 'tree of life' devised by evolutionists during the nineteenth century. Aristotle clearly appreciated the great divisions of the animal kingdom, and even understood the mammalian nature of whales and dolphins. *Historia Animalium* ran to several volumes, and was, for the period, a remarkable compendium of information, much of which was completely accurate.

Aristotle appreciated the habits of animals from the point of view of the animals themselves. That in itself was a revolutionary attitude towards nature. The Greek scholar's observations on fish behaviour, especially on the behaviour of the species of catfish which lived in the Achelous river, were excellent. His accounts of fish anatomy were not matched until the English parson, John Ray, studied the subject in the seventeenth century. Aristotle also made a thorough investigation into the biology and behaviour of octopuses and their allies, squid and paper argonauts. His writing is littered with cameos of behaviour. For example, he records how woodpeckers place nuts in crevices so that they can crack the shells more easily. And yet Aristotle fervently believed that swallows hibernated during the winter, and that snakes had an insatiable thirst for wine!

Aristotle's writings could have laid a secure foundation for biological science. Unfortunately, the spirit of free inquiry which he displayed languished for the next two thousand years. Even the wisdom embodied in his encyclopedia suffered, as yarns, tall stories, and superstitions were added to it indiscriminately. Pliny the Elder, a Roman civil servant and cavalry officer living three and a half centuries after Aristotle's death, was a major offender. He was a collector of stories related by sailors, farmers, and the common people. These he incorporated into Aristotle's work, extending it to thirty-seven volumes. It appeared as Pliny's *Historia Naturalis* (Natural History) and was a ramshackle amalgam of facts and fancies. Pliny seemed to believe everything he was told. Thus he reported that there were countries in Africa ruled by dog-headed people; he also maintained that huge snakes hid in rivers to snare elephants as they drank. Through his influence, superstitions became accepted as facts, imaginary tales passed as true. Pliny also originated the notion that every animal and plant had a 'use'. He suggested that none had lives of their own, but were intended solely for the benefit of mankind. This is a view which has its adherents even today.

Unfortunately, Pliny's natural history was widely copied. It formed the basis of the *Physiologus* – The Natural Historian – which was probably written in Alexandria during the second or third century A.D. This somewhat unreliable book became the standard reference work for all those with a curiosity about nature. It also consigned the study of animals into the realms of fantasy for the next fourteen hundred years. The *Physiologus* served to reinforce all kinds of strange ideas about animal behaviour.

During the Dark Ages, explanations for animal behaviour were shaped in ignorance. As alchemists had implicit faith in their ability to transmute base metals into gold, ordinary

Early writers contributed to the belief in fabulous monsters

The origin of a myth. A female crocodile carrying her young in her mouth – not eating them

people believed that animals had a magical capacity to change from one kind into another – barnacles into geese, for instance. The origin of some species was attributed to a process of spontaneous generation. How else could maggots seem to arise from the flesh of rotting meat, and fleas from dirt? People also generally believed that mice were somehow spirited out of old rags, and that crocodiles were formed from the slime and mud on the edges of rivers.

Animal behaviour was at the root of many superstitions. The nocturnal hooting of tawny owls during the autumn was not interpreted as choral duelling between birds sorting out territorial boundaries, but as death's dread messengers prophesying doom. Similarly, the laughing call of the green woodpecker was assumed to signify that rain was on the way. Some old-fashioned beliefs along these lines persist to this day. In a few areas, for example, the green woodpecker goes by the utterly inappropriate name of 'storm cock'. There are country folk still who cherish the notion that it occasionally rains frogs and toads. In fact, these creatures regularly make overland journeys, but avoid doing so in hot, dry conditions. When the countryside has been drenched, then they may disperse from ponds and so appear in large numbers in fields and village streets. Until very recently, many people – even scientists – believed that crocodiles cannibalised their young. This idea was based upon a complete misinterpretation of the reptile's behaviour. In fact, mother crocodiles take their hatchlings into their mouths when transferring them from breeding sites to nursery areas.

During the eleventh century, the *Physiologus* was given a fresh lease of life by Christian monks who applied their scholarship and writing skills to recording in detail what was known about God's world. This development coincided with a rediscovery of the ancient texts. Animals began to occupy the minds of the devout; zoos appeared in stained glass, and menageries were carved in stone on the sides of cathedrals.

It was not altogether surprising that the Church should foster an interest in the natural world. After all, the Bible was very explicit on the subject of how and why animals and plants were made. Sooner or later, Christians were bound to feel obliged to acquaint themselves with the fruits of God's creation. However, at first, they had an unusual approach to nature study; rather than watch animals in action, they preferred to consult the books of the ancient Greek and Roman scholars who, they believed, had discovered everything! Foremost among those books was the *Physiologus*. It was as close to an inventory of God's creatures as the monks in their seclusion could obtain. As part of their devotional duties, they made copies of it to which they added exquisite illustrations. They called them bestiaries.

The bestiary was immensely popular during the Middle Ages, and described the weird and magical world which appealed to medieval taste. Although early bestiaries were close to the original *Physiologus* in content, succeeding generations of monks felt free to add to the text travellers' tales and still more superstitions. Although bestiaries became larger and more colourful, they also became increasingly unreliable as sources of information.

The bestiary was intended as a holy text, a scrap-book of God's handiwork. Many were wonderfully illuminated with gold because only this precious metal was good enough to decorate an account of His creatures. Several hundred have survived, and an especially lovely one, dating from between 1150 and 1200, is preserved in the Bodleian Library, Oxford. As one turns the stiff vellum pages, the dedication and supreme craftsmanship of the authors and artists are revealed by the quality of the work: expanses of burnished gold dazzle the reader, and the colours, though eight hundred years old, are so vivid that they could have been applied yesterday.

The bestiary starts by depicting God creating the heavens and earth, and naming the beasts of the field, the fowls of the air, and the fish of the sea. Thereafter the text is adorned with charming cameos of accurately depicted behaviour: a white stork is caught in the act of seizing a frog; a cat eyes a mouse suspiciously while its companion grooms, delving into its hind quarters in a characteristically feline manner; a sea eagle carries in its talons a fish, head to wind like a torpedo; magpies mercilessly mob an eagle owl; a peacock is portrayed in all its finery. The monks may well have been familiar with all of these species. However, they had a little trouble illustrating many kinds of creatures with which emphatically they were not acquainted.

Monks were arm-chair naturalists, and their art-work shows how they wrestled with descriptions of animals provided by explorers of far-away places. They illustrated the whale as a huge fish, with gills, and not as a mammal. Their crocodile has the appearance of a spikey-backed dog, although the artist got its mouth more or less correct, equipping it with gums bristling with daggers. If the monks were a little sceptical about the accounts which were related to them, then perhaps the travellers themselves may have hardly believed what they saw. What could be stranger than a bird carrying a bag beneath its beak – the pelican? Or more extraordinary than a 'fish' with an ivory lance two metres long growing from its snout – a narwhal? Perhaps such apparent impossibilities lay at the root of some of the fabulous animals documented in the bestiaries. Europeans observed for the first time fantastic horned creatures such as the rhinoceros and the oryx; Viking seafarers caught tantalising glimpses of narwhals, and perhaps came into possession of their gleaming twisted tusks. The unicorn may have been derived from a melange of these species, together with the cloven feet and beard of a goat grafted onto the body of a horse.

It is more difficult to trace the origin of the terrifying and thoroughly unpleasant animals which were thought to lurk in the medieval countryside. It paid people to tread carefully when a basilisk was around because its breath could shrivel birds on the wing, and its gaze alone was lethal. Dragons were no more friendly. The bestiary confirms the existence of several sorts of fire-breathing dragons. Luckily, they were comparatively rare. Although the native English dragons were normally reclusive in habit, living the life of smouldering hermits, in places flights of these fractious serpents were more common than birds – four hundred were reported in one flock! Dragons apparently had a prodigious appetite for sheep and unwary people. But their fondness for elephants was suicidal: a dragon was able to snare an elephant by lassoing its legs with its tail, causing the elephant to fall and crush its attacker. People throughout the world were bothered by the annoying habits of dragons. The Chinese variety occasionally took to feeding on the sun, causing the well-known phenomenon of the solar eclipse. When this happened, people ran into the streets, beating drums to frighten the hungry creature away – a course of action that was invariably successful! The bestiary also records the fact that dragons were the traditional sport of brave knights and valiant saints. St George gained universal acclaim for his heroic slaying of a particularly notorious beast that terrorised the city of Silene, in Libya. Serious doubts were expressed as early as 1481 about the existence of dragons, but it was not until the end of the seventeenth century that these monsters were deemed to belch fire and spread havoc only in the imagination.

And yet not all myths are necessarily untrue. A myth may

Salamandra uocata qͥ contra incendia ualeat.
cuius inter omnia ueneuata uis maxima est; ce
tera enim singulos feriunt. hec plurimos pariter interi
mit. Nam ĩ si arepserit omnia poma inficit ueneno.
̉eos qui ederint occidit; Qͥ ́ etiam uł si in puteum ca

be a description of something in one kind of language described today in quite another. For example, the bestiary faithfully mentions and illustrates the mermaid and merman. Sea monsters were, without doubt, observed by ancient Norse mariners as they forged their way across the northern seas. The huge creatures loomed out of the water. Each had the shoulders of a man but no arms or hands, and appeared mystifyingly slender at its lower end, where the body met the sea. Although such descriptions were treated lightly by zoologists, they are nevertheless perfectly accurate accounts of marine mirages of seals or walruses inquisitively peering out of the sea. The distorted images would be observed most dramatically from the decks of low-slung boats such as those used by the Vikings. The number of reports of marine monsters, and the sizes they attained, tended to decrease as ships became taller! Newts crawling from hearths may also have inspired the story of the mythical salamander which emerges from fire. Real-life salamanders are strikingly marked black and yellow amphibians which, in central Europe, often hibernate in hollow logs. When the wood is thrown onto open fires, the heat causes the animals to stir from their winter torpor and, to the astonishment of onlookers, try to escape from the flames.

In the days when the bestiaries were fashionable reading matter, the text was designed to reinforce the view that the universe was governed by God, that He made the whole world in the service of mankind, and that everything concealed a message for the eyes of the faithful.

Every creature in the world
A book and picture is for us
And like a mirror too.

Such a view may well have been a legacy from Pliny!

In the Middle Ages, Christian morality pervaded thoughts and deeds, and so the moral significance of animals was studied more intensively than were their physical forms or habits. All sorts of Divine Truths were revealed: monkeys and apes were made for us to laugh at; lions and tigers were created to give us a healthy sense of fear; and should we ever become too arrogant, then there were always fleas and lice around to make life a misery.

But the bestiary had more serious lessons to teach. The elephant's massive bulk was taken to be a demonstration of the might and power of God. The bestiary also claims that these animals have no desire to mate – at least on Earth. This peculiar

Salamanders emerging from fire, from a bestiary of *c.* 1250. Our ancestors were astonished when creatures like European or fire salamanders (above) crawled out of burning logs

The glance of a basilisk was
thought to be lethal to both
people and birds

Mistaken observations by
seafarers led to a belief in
fearsome sea monsters

piece of natural history is based upon the fact that elephants copulate so infrequently that travellers would have rarely caught them *in flagrante delicto*. The pious chroniclers therefore seized upon the elephant couple's apparent lack of libido and caused it to represent all that was virtuous in Adam and Eve before they were tempted to chew on the apple and become informed of things like sex. Elephants were thought to make the journey to Paradise in order to make pachydermous love, and this monumental event was followed two years later by the birth of a calf. Even if the monks managed to record the gestation period more or less correctly, they certainly underestimated the elephant's carnal knowledge!

Of the lioness, we are told that she brings forth her cubs dead, and watches over them for three days and nights until the male arrives, breathes into their faces, and so generates life in their little bodies. This piece of improbable behaviour was supposed to bring comfort to all good Christians, by reminding them that God the Father raised Jesus from the dead on the third day after his crucifixion.

Even the habits of mythical animals had theological lessons to drive home to simple churchgoers. The images of terrible dragons glowering from the pages of the bestiary, or from stone porticoes, reminded the readers or worshippers of the power of the Devil, and of the Church's vision of Hell. The bestiary interprets the meeting between the unicorn and a virgin in the forest as an allegory of Christ and the Virgin Mary. When the fearsome little animal perceives that the girl is completely virtuous, it approaches her meekly and lays its horn in her lap, and so can be slain by a knight errant. The scene is symbolic of the passion and incarnation of Christ.

Perhaps the most curious piece of fabulous behaviour recorded in the bestiary was that of the Phoenix. According to tradition, only one Phoenix at a time could live in our world. Its true home was Paradise, a land of unimaginable beauty lying beyond the distant horizon towards the rising sun. Nothing dies in Paradise, and here was the crux of the bird's dilemma. After a thousand years had passed, the Phoenix became oppressed by the burden of its age; the time had come for it to die. To do so, the Phoenix had to wing its way into the mortal world, flying westwards across the jungles of Burma, and the torrid plains of India, until it reached the scented spice groves of Arabia. Here it collected a bunch of aromatic herbs before setting course for the coast of Phoenicia in Syria. In the topmost branches of a palm tree, the Phoenix constructed a

nest out of the herbs, and awaited the coming of the new dawn which would herald its death. As the sun soared above the horizon, the Phoenix faced east, opened its bill, and sang such a bewitching song that even the sun god himself paused for a moment in his chariot. After listening to the sweet tones, he whipped his horses into motion and a spark from their hooves descended onto the Phoenix's nest and caused it to blaze. Thus the Phoenix's thousand-year life ended in conflagration. But in the ashes of the funeral pyre, a tiny worm stirred. Within three days, the creature grew into a brand-new Phoenix, which then spread its wings and flew east to the gates of Paradise in the company of a retinue of birds. The symbolism is not too difficult to understand. The Phoenix represents the sun itself, which dies at the end of each day, but is reborn the following dawn. Christianity took the bird over, and the authors of the bestiaries equated it with Christ, who was put to death, but who rose again.

During this period there were few people who were capable of divorcing themselves from the medieval attitude to see for themselves how animals behaved. One such was Frederick II of Hohenstaufen, Holy Roman Emperor, King of Sicily and Jerusalem. Unlike most of his thirteenth-century contemporaries, he was a brilliant and highly literate man who was profoundly interested in the countryside. His fascination for animal habits stemmed from his preoccupation with falconry.

Falconry is an ancient sport which probably evolved on the wide plains of Turkestan as long ago as three thousand years. It demands a deep understanding of hawk behaviour and the habits of the birds at which falcons are flown. Although the sport was practised in Europe by the Germanic tribes between A.D. 100 and 500, there is evidence that it became popular in a widespread fashion as a result of the Crusades. Princes and knights made contact with middle-eastern potentates who were themselves devotees of falconry, 'the noblest of arts'. This increase in popularity coincided with the reign of Frederick II.

During the early Middle Ages, Frederick II was one of the most enlightened and educated monarchs in Europe. Apart from his continual involvement with politics, he pursued with enormous success interests in mathematics, architecture, and natural history. He even founded the University of Naples in 1224. But his great passion was falconry. His dedication to the sport was such that he once abandoned a siege because he insisted on taking a day off at a rather critical time in order to

A page from Frederick II's *De Arte Venandi Cum Avibus*, showing migrating geese and white storks. Above, white storks migrating from Europe to Africa over the Bosporus

de remotioribus locis septentrionalibus, habent transire per eo quod apud ipsas citius incipiat frigus propter longum iter quod habent facere citius properant ad recedendum. Ille vero eiusdem speciei que a minus remotis locis hoc transire quin tardius adveniet eis frigus et quod breve iter habent facere tardius incipiunt transire. Amplius si tempus autumpni sint april et venti faventes sine intermissione transibunt et citissimo complebunt transitum suum. Si fuerit septum ut frigus inveniat et venti non faveant tardius complebunt transitum suum expectabunt enim semper quod illa temporis asperitas pretranseat et aves quanto magis sint prope equinoctialem tanto tardius transire incipiunt. Ordinem autem huiusmodi fiunt in transeundo. Omnes enim ille aves que de rivera dicuntur non confuse volant nec inordinate. Quadmodum volant ille que terrestres sunt. Nam terrestres non attendunt in suis volatibus que prior sit in illa multitudine aut que posterior. Iste vero de rivera ad plus huiusmodi ordinem servant. una precedit alie omnes de illa

multitudine subsequuntur successive et in gemina serie una series erit a dextris alia a sinistris et adherescunt plures in una serie quam in alia que apparuerunt precedenti uni ad modum linearum concurrentium ad faciendum quendam angulum et aliquando in una serie solum et istum ordinem fiunt non solum in transeundo ad loca remota et redeundo de illis verumetiam in eundo ad pascua et in redeundo de pascuis ut predictum est. Una autem precedit sepe maxime enim in multitudine gruum non pro eo quod illa sciat loca ad que iture sunt et alie non. Sed ut prevideat nocumenta et alias reddat vel abstinendo aut deviando cautiores. Reliqui namque in tutela illius que precedit que dicitur dux asserantur in volatu suo. Et quin labor est illi ducum precedere non solum per volatum. Sed propter sollicitudinem quam habet et propter timorem erit de ducatu suo et de ordine. Et si amplius huiusmodi laborem tollerare non potest et alia succedit ei in ducatu. Qui vero erunt reintrat ordinem aliarum. Non erit ergo veri

fly his falcons! He practised his hobby mostly in the spacious countryside of Apulia, Calabria, and Campania in southern Italy, where he built dozens of castles and hunting lodges.

It was his scholarship that made Frederick II such a luminary in the constellation of medieval monarchs. He studied not only his hawks but took the trouble to investigate birds in general, because the information so gained gave him a greater mastery over the sport. Above all, he recorded the results, both of observation and speculation, in an immense and profusely illustrated 'book'. He called it *De Arte Venandi Cum Avibus* (On the Art of Hunting with Birds) and completed the epic work in around 1250.

When one examines the volume preserved in the Vatican Library, the depth of Frederick's insight into bird behaviour is readily apparent. He had no time for Aristotle's assertion that birds hibernate in winter; he knew from his own experience and inquiries that many species make two excursions each year, travelling south during the autumn and flying north during the spring. He heard the calls of migrating cranes, herons, and geese passing overhead during the night 'talking to their fellows'. As if to add force to his ideas about migration, the Latin text is illustrated with formations of white storks winging their way purposefully across the pages. He recognised the apparently compulsive nature of the migratory urge. 'No matter how bad the weather, and in spite of fatigue, birds persist in migrating; once they start, they carry on as though it was the most important object in their lives.' No one would argue with that statement today.

His knowledge spanned the whole of ornithology. He meticulously recorded the various different styles of flight, from the dashing movement of the scimitar-winged swift to the lumbering motion of swans. He was well aware of the role of the preen gland among aquatic birds. Astonishingly, he discovered the essential details of the hen cuckoo's breeding behaviour. He correctly stated that she neither builds a nest nor feeds her young, but deposits her eggs directly into the nests of other birds. Frederick verified this by hand-rearing an odd-looking chick from a song bird's nest, and it grew into a cuckoo. Although Frederick was right in what he wrote some seven hundred and fifty years ago, controversy over the laying behaviour of cuckoos continued to rage until the twentieth century; some naturalists staunchly maintained that the hen cuckoo laid first on the ground and then transferred her eggs in her bill to the foster parents' nests.

A cuckoo about to lay into a reed warbler's nest, removing one of the warbler's eggs. Frederick II described this behaviour accurately in the thirteenth century, but birdwatchers disputed it until the twentieth

Frederick II's book is also a practical manual on falconry, and sets out in detail how to train birds to the peak of perfection. The process includes the use of a lure which acts as substitute prey; the falconer swings it through the air on the end of a line to trigger the free-flying falcon into a deadly stoop. The lure is reduced to the bare essentials of a flying bird, a wing and movement – analogous to a rough duck decoy or a fisherman's fly. Falconers like Frederick therefore inadvertently discovered and exploited an important principle of animal behaviour, namely that animals often react only to a narrow band of features in their surroundings.

The royal naturalist insisted on checking the veracity of many of the more outlandish beliefs of his time. One stated that a northern species of wildfowl – the barnacle goose – whose nesting haunts were unknown, was generated not from an egg but from a worm. Frederick had heard that in the far north old ships were to be found in whose rotting hulls worms were born which magically transformed into geese. Each goose hung from the dead wood by its beak until it was old enough to fly. This improbable story had such wide currency that Frederick felt obliged to investigate the matter, and dispatched an envoy with orders to bring back some of the miraculous timber. The mission was successful. The messenger returned with samples of wood encrusted with 'goose barnacles' which bore only a superficial resemblance to birds. Frederick correctly concluded that the superstition arose from the migrating habits of the geese, which vanished during the summer to some remote latitude. As a result, men, in ignorance of their real nesting location, had invented a fanciful explanation for the annual disappearance of the skeins of cackling geese. Nevertheless, the story was not laid to rest for several centuries, thus justifying the pious in devouring geese on Fridays, along with the less appetising fish sanctioned by the scriptures!

Frederick II might have brought about a renaissance in zoology and his writings could have inspired in the minds of generations a keen interest in animal behaviour. Unfortunately, he was centuries in advance of his time in the quality and originality of his observations. *De Arte Venandi Cum Avibus* was not widely read; it was compiled before the age of printing, and the laborious copying methods of the time ensured that only a few copies were produced. To an overwhelming extent, therefore, the wisdom encapsulated in *On the Art of Hunting with Birds* went to waste. Frederick II therefore goes down in history as an isolated phenomenon.

A real-life barnacle tree – an upturned root covered in goose barnacles – and the barnacle tree of the Middle Ages, showing the transformation of barnacles into geese (below right). Above, the barnacle goose itself

It was left to those who compiled the bestiaries to influence the course of natural history. The original beautiful hand-copied books were extremely rare, and zealously guarded in the monastery libraries. But with the advent of printing in the latter part of the Middle Ages, duplicated versions of the bestiaries became widely available. Information about animal behaviour was also incorporated into folk-tales and fables, and these too were committed to the printed page during the late fifteenth century. Like the bestiaries, the stories all had a strong moral purpose, clarifying virtues and elucidating expectations.

They also used animals to tell the truth about people. The fable of the ant and the cicada provided a cautionary lesson for every peasant community. The story relates how the pleasure-seeking cicada whiles away the summer in song, while the industrious and far-seeing ant slaves away to stock his under-ground larder. When winter comes, the insect pays for its carefree attitude towards life by starving to death, whereas the hard-working ant profits from his labour and so survives the cruel, cold, lean months.

In the fables, the animals are used as convenient metaphors for human behaviour, and the fables' success depends upon the fact that their characters are based upon a general re-cognition of biological reality. Everyone knew that cicadas trill throughout the balmy months of summer while ants beaver away, and that cicadas die during the winter, while ant colonies thrive until spring on their stores.

All countrymen understood the nature of the fox, and this was crucial to the enormous success of the folk-tales of Reynard the Fox which were widely circulated during the Middle Ages. Among the first pieces of literature to be printed, they were scurrilous stories about village politics, full of ribaldry, trickery and skulduggery, in which the animals put on a charade of human cupidity. Reynard was the hero, and used his own sense of native cunning to keep one step ahead of officials and establishment figures all too familiar to village folk.

Casting animals allowed each character to be endowed with instant personality without any need to elaborate. The fox was chosen as the focus of the village lampoon because the readers readily grasped the role that the fox played in the countryside. It was a lone hunter, a poacher, a breaker of conventions, appearing half dog and half cat, at home both by day and night in the woods and hedgerows. By using a fox as a champion of

The cunning nature of the fox was understood by everyone in the Middle Ages and formed the basis of countless fables, including those of Reynard (above)

peasant aspirations, a whole galaxy of qualities could be taken for granted in the hero's character.

This convention is still employed in childrens' books, even the most modern.

Towards the end of the Middle Ages, however, there was a rejuvenation of the open-minded approach to nature pioneered by Aristotle. The new approach was being fostered by people like Nicolaus Copernicus. In 1543, he helped to launch Europe into the Renaissance by daring to publish the blasphemous theory that the Sun and not the Earth was the centre of the solar system, thus refuting the Biblical version of planetary geography. This heralded the beginning of a renewed curiosity about the nature of the world and its constituents. Galileo laid the foundation for mechanics; William Gilbert provided the basis for experimental physics; and William Harvey discovered the secrets of the circulation of the blood. Such men made progress because they were sceptics, able to free their minds from medieval prejudice. Knowledge, like the universe, was proving to be limitless!

As the fresh intellectual climate spread across Europe, it was only a matter of time before even clerics, whose minds had been steeped in folk tales and bestiaries, became imbued with the spirit of scientific inquiry. They began to question the truth of their texts, and sought to verify for themselves how animals acted in their natural surroundings. Their discoveries proved that God's world was even more marvellous than they had previously suspected, and at last took the study of animal behaviour beyond the realm of fantasy.

Children's books, even the most modern, employ the age-old convention of using an animal's general behaviour to create an instant personality

ONCE upon a time there was a frog called Mr. Jeremy Fisher; he lived in a little damp house amongst the buttercups at the edge of a pond.

IN PRAISE OF GOD

John Ray (1627–1705)

In 1649, Charles I was executed. In that same year, a theological student called John Ray was awarded a Fellowship at Cambridge University. He was to become seriously interested in natural history, observing and recording animals and their habits. The relationship between the Stuart king's execution in London and Ray's increasing absorption with nature was more than simple coincidence. For the previous eight years, England had been racked by civil war. The bitter political and religious rivalry between the factions drove many people to retreat to the comfort of nature, which appeared, by comparison with the affairs of men, beautiful and wholesome. Some such consideration seems to have influenced Ray.

John Ray came from humble stock. His parents were neither educated nor rich. His father was the blacksmith in Black Notley, a north Essex village, and his mother was the local herbalist. John Ray was born in the smithy on 29 November 1627. By all accounts, he was a brilliant boy, and influenced by his parents' work. He accompanied his mother on her rounds of the fields and lanes collecting the plants she needed to dispense to the sick. The rambles encouraged in him an interest in botany. As a young boy, he also spent hours in the forge, watching his father hammering out horse-shoes and plough-shares. Perhaps this experience encouraged in him a trait that he was later to apply to the details of floral and animal anatomy: an intense desire to understand how things worked.

The part of East Anglia where he lived was an excellent place in which to indulge a liking for natural history. Elms presided over the deep lanes which skirted the parish boundaries. Robins and wrens nested in their banks among mats

God naming the animals. According to the scriptures, the birds and sea creatures were made on the fifth day of Creation and the animals on the sixth

of stitchworts and clusters of campions. In summer, sheep's parsley dressed the hedges in a lacy haze. Nightingales made music in cowslip-carpeted copses, and lapwings nested in meadows bright with purple fritillaries. A few miles to the east of Black Notley lay a totally different world, a world of saltmarshes and muddy creeks, where damp, grey, mists clung to the ground. Then as now it was the home of myriads of waders, flocks of which would emerge from the swirling mist, to vanish in an instant. Here Ray could listen to the piping of oystercatchers, the mournful bubble of curlews, and the merry trilling of redshanks, locally known as wardens of the marshes. Out beyond the saltings, skeins of brent geese and wigeon followed the ebbing tide to gorge themselves on swathes of eel grass. A short journey to the north of his home took Ray into the fens, with their expanses of bogs, their lakes, and the vast reed beds in which bitterns lurked while marsh harriers wheeled overhead. To the west, on the border between Essex and Cambridgeshire, were chalk pastures studded with orchids, and frequented by butterflies with wings of intense blue.

By good luck, Ray's talents were recognised by two local clergymen, and through the influence of one of them, he was accepted by Cambridge University at sixteen years of age. During the period of the Civil War, he buried himself in his studies, and was recognised for his outstanding achievements in the year that Oliver Cromwell abolished the monarchy. He embarked upon an academic career, taking up successive lecturing appointments in Greek, Mathematics, and the Humanities. However, he was increasingly attracted to the Church, as were most educated people.

In 1660, the monarchy was restored, and Ray took Holy Orders. This gave him a professional interest in the work of the Creator. Like all clerics at the time, he believed that every animal and plant was created to reveal some aspect of the Divine Plan. A study of the works of nature was widely recommended as a key to the understanding of His mind.

'Nature is of God; its study is His service; its truth, His wisdom.' Studying nature led to nature's God. But Parson Ray was different from the majority, in that his mind was unclouded by the mists of medieval myths and magic. This enabled him to watch toads spawning in ponds without thinking of witches, and to observe grass snakes gliding through the buttercups without meditating on the deeds of the Devil. Although the Church encouraged people to approach the living world with a sense of wonder, it acted as a powerful deterrent to those who were tempted to cultivate a scientific attitude towards 'natural knowledge'. The Bible implied that God had made the Universe in perfect order. It was considered impertinent to inquire too closely into how things were designed and how they worked. Unusually, Ray did not take this stricture too seriously. Indeed he later wrote of his immense 'gratitude to God' that he should be born in an 'age of noble discovery', when the secrets of nature were being 'unsealed and explored'; he was referring to such revelations as the ceaseless circulation of the blood, the weight and 'elasticity' of the air, and the structure of the organs of generation. But Ray noted with scorn that there were those who condemned the passion for knowledge, and the study of experimental philosophy as pursuits unpleasing to God, 'as if Almighty God were jealous of the knowledge of men'. After all, such people should remember that knowledge made us 'superior to animals and lower than the angels, capable of virtue and of happiness such as animals and the irrational cannot attain'.

It certainly was an age of change. With the Restoration, 'experimental philosophy' – or science – flourished and the grip of the Church on the minds of men began to weaken. Charles II himself took a keen interest in scientific matters, and in 1662 granted a royal charter to a society devoted to promoting scientific experimentation – the Royal Society. Paradoxically, that was the year in which John Ray's university ambitions were shattered by the Act of Uniformity, which demanded his assent to the Thirty-nine Articles of the Anglican Prayer Book. Ray was of the Puritan persuasion, and refused to swear the necessary oath, and so he, together with four thousand other clerics, was summarily denied a living. As it happened, the Act did natural history a favour. If Ray could not serve God as a preacher, he would do so by immersing himself in natural history and writing about it.

He had already encountered one major problem. The countryside was a wonderland of treasures lovingly and exquisitely produced by God. Yet many of the wayside flowers had no names, the secretive brown birds which stuttered and churred from stands of reeds were referred to by different names in every parish, and insects were a complete mystery. Ray identified the crux of the problem: 'When men do not know the names and properties of natural objects and are ready to believe any fanciful superstition about them, they cannot see and record accurately.' The books available were of little assistance: the herbals were incomplete and often unreliable,

A great auk, now extinct, from Willughby's *Ornithologia*

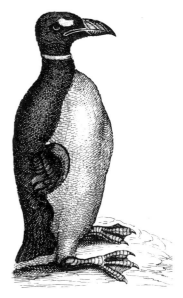

and the bestiaries, even worse! Ray resolved that a catalogue was required – a *Systema Naturae*. He therefore planned a monumental inventory of animals and plants which would put an end to the confusion and enable people to identify what they were looking at. It would include descriptions of each species, its breeding habits, and, where appropriate, accounts of its food and behaviour. The work would be a celebration of God's creation.

Various attempts had been made in the past to assign species to their rightful places, but most had been based upon meaningless criteria: Aristotle had classified animals according to their size, so that elephants and ostriches were the first mammals and birds respectively; one medieval method placed living creatures in alphabetical order. For the first time, Ray grouped animals and plants on the basis of anatomical similarity, anticipating the practice of Carl von Linné, the Swedish taxonomist whose system is used today.

Ray was assisted in his grand venture by one of his students, the Hon. (later Sir) Francis Willughby, a man whose vitality inspired the more cautious Ray. When the Act of Uniformity left Ray unemployed, his younger friend became his patron, and financed extensive field trips.

Travelling in the seventeenth century was an uncomfortable and disagreeable business and required true grit. The inns were overcrowded and malodorous; the coaches were draughty and rattled the occupants about like peas in a pod; and in addition to bed-bugs and saddle soreness, there was the ever-present danger of robbery.

Despite these hazards both Ray and Willughby explored the length and breadth of the kingdom, collecting plants and watching animals wherever they went. They picked saxifrages and stonecrops on the slopes of Snowdon. On the north-east coast they visited Holy Island, and observed seabirds on the Farne Islands. They saw puffins – their comical, poster-coloured beaks crammed with sand eels – nesting in burrows among swards of sea pinks. They watched guillemots jostling for body space on their breeding ledges; heard the courtship whimpers of eider ducks; and marvelled at the tameness of nesting shags and kittiwakes. As they walked over the Farnes, they were attacked by sea swallows which treated the pair as marauding predators. Willughby and Ray came to the conclusion that these dainty but vicious birds should be ranked with the sea-gulls and not with the swallows which twittered around village ponds during the summer. They also realised that blackcock

A puffin from Willughby's *Ornithologia*

A puffin returning with food for its chicks. John Ray observed these in the Farne Islands and assigned them to the auk family

and greyhen were the male and female of the black grouse, not two separate species.

In 1663, Ray decided to make a three-year grand tour of Europe, starting in the Low Countries, wending his way through Germany towards Vienna, and then to Venice. In Padua, he studied anatomy, and in the markets of Rome he recorded copious details of the fish and fowl on display; for good measure he climb Mount Etna.

These and other journeys, often in the company of Willughby, broadened his mind with a wealth of first-hand experiences denied to closeted monks and arm-chair philosophers. For example, people still believed that swallows and swifts hibernated in hollow trees during the winter, or buried themselves in the bottom of ponds. Ray and Willughby were fairly certain that this theory was not true. Ray had himself witnessed these birds heading out to sea during the autumn. An observation of this kind was one of the benefits of getting out into the field.

Sir Francis Willughby died in 1672, leaving only a small pension to his former tutor. Ray returned to his native Essex village where he continued to serve God for the last twenty years of his life by writing. He probably wrote most of Willughby's *Ornithologia*, published posthumously in 1678. It was a massive illustrated handbook of birds, packed with information, and its publication launched scientific ornithology. In its pages was buried the most significant discovery relating to behaviour made during the seventeenth century – the territorial nature of birds. Ray had read a book about bird lore by G. P. Olina, published in Rome in 1622. Of the nightingale, Olina wrote that 'it sings in its freehold'. In *Ornithologia*, Ray enlarged on this theme. 'It is proper to this bird at its first coming to occupy and seize upon one place as its freehold, into which it will not admit any other nightingale but its mate.' Aggressive rivalry among birds in the spring had been well-known since the days of Aristotle, but no one before Ray had connected this behavior with the claim of cock birds to exclusive occupation of breeding areas.

Ray wrote subsequent volumes on fishes, mammals, and plants. Basilisks, dragons, and Phoenixes were banished from these works of true scientific scholarship.

Ray saw in the variety of the animal world, in form and behaviour, a demonstration of Divine power. In His wisdom, God had given His creatures different means of preserving themselves. Ray recorded the fact that some, like rabbits, 'dug vaults to secure themselves and their young'. Others, like

Luscinia
The Nightingale.

A cock nightingale proclaiming his territory by singing. John Ray, in Willughby's *Ornithologia*, was one of the first to understand the territorial nature of birds. Above, a nightingale from Willughby's book

hedgehogs, were armed with defensive prickles. Those which had no armour tended to be fast-moving. He noticed, too, that the hare even had eyes set on top of its head so that it could see all the way round and have 'the enemy always in its eye'. The long mobile ears seemed designed to be extremely sensitive so that the hare 'be not suddenly surprised or taken napping'. The insatiable appetite for copulation which both animals and humans displayed was 'part of the Great Design of Providence' to ensure the continuation of every species. Ray also considered the question of why birds should lay eggs rather than bear live young. To him, it was manifest evidence of the wisdom of God; if birds had been viviparous, 'the burden of their womb would have been so great that their wings would have failed them'.

Parson Ray was not averse to experimentation; indeed, his scientific approach to the secrets of nature gained him admittance to the Royal Society in 1667. He established the relationship between caterpillars, chrysalises, and butterflies by keeping grubs which fed upon his vegetables and watching them transform, stage by stage, into cabbage whites. He took an egg daily from the nest of a swallow, and caused the poor bird to lay a total of nineteen eggs before it gave up and deserted – the normal clutch for a swallow is five or six – and reasoned that birds cease to lay after they have produced a sufficient number of eggs to cover.

Ray was one of the first people to write about 'instincts', which he defined as behaviour in an animal directed to 'ends unknown'. Needless to say, he saw instincts as the work of a thoughtful God. He observed the way nestlings, their eyes tight shut, elevate their rumps high in the air to void their excreta away from the nest, and so help their parents seize and remove the waste as it emerges. He studied birds' nests and noted that each species appeared to have its favourite kind of construction. Long-tailed tits wove incredibly intricate spheres of lichens, mosses and spiders' silk; song thrushes smeared their grass cups with mud. He observed an amusing example of instincts taking their natural course in the barn-yard: when a hen fostering a brood of ducklings brought her brood to the brink of a pond, the baby birds headed for the water 'tho they never saw any such thing before, and tho the hen clucks and calls and doth what she can to keep them out'.

Ray's real scientific genius was revealed in 1691 when his most widely read book was published: *The Wisdom of God Manifested in the Works of Creation*. In it he tackled the delicate problem of fossils, which were thought to be the remains of

Designed for survival. The hare's eyes, set on the top of its head, and its long, dished ears, ensure that it is not easily surprised or caught napping

creatures drowned in the Flood. The opening verses of Genesis were precise in stating that God populated the world with a fixed number of species during the hectic week of Creation. Representatives of all of them were taken aboard the Ark by Noah during the Deluge.

From the evidence, Ray doubted the veracity of this version of events. He had found a great variety of fossils in very strange places. He knew that some occurred on the tops of mountains. Biblical mathematics suggested that the Deluge lasted ten months. That seemed too short a period for even the most torrential rain imaginable to have swamped mountains. Ray was also puzzled by strange fossils such as 'Devil's toe-nails' (brachiopods) and 'thunderbolts' (belemnites). These and the ammonites he found at Whitby and elsewhere were clearly the remains of creatures which no longer lived on planet Earth.

To Ray, these facts were a source of considerable confusion. As a scientist, he had to accept the evidence that many kinds of animals were now missing from the world. But the Church, and Ray was a deeply religious man, was unwilling to concede such blasphemy because the 'destruction of any one species meant a dismembering of the Universe, thereby rendering it imperfect'. The discovery of fossils, on the other hand, indicated quite clearly that God had failed to maintain a perfect world!

Ray the scientist cast doubts upon the religious explanation of nature which would not be resolved before Darwin's masterly synthesis, nearly one hundred and fifty years later. Meanwhile, Ray's animals and plants were considered to be immutable, isolated acts of Creation, quite separate from mankind. Yet, by the time Ray died in 1705, the study of nature was beginning to cast increasing doubt on this view.

Whales and dolphins had become firmly classed as mammals, undermining the accuracy of the scriptural account of Jonah being swallowed by a monstrous fish. Starfish proved to be animals and not drowned celestial bodies; sea anemones had turned out to be ravenous sea creatures, not decorative marine flowers; and corals to be animal polyps, not aquatic plants. People like Robert Hooke, the curator of instruments at the Royal Society, and Anton von Leeuwenhoek, a draper from Delft in Holland, were now subjecting nature to examination through microscopes. Whole new worlds teeming with marvellous little organisms came to light through the lenses of their instruments. Here they saw animals that resembled the brutes of the myths; a puny green fresh-water polyp barely visible to

The multi-headed Hydra, a small freshwater polyp, named after the Greek monster that grew two new heads whenever one of its own was cut off. Below, the microscope reveals a fascinating world teeming with minute strange organisms

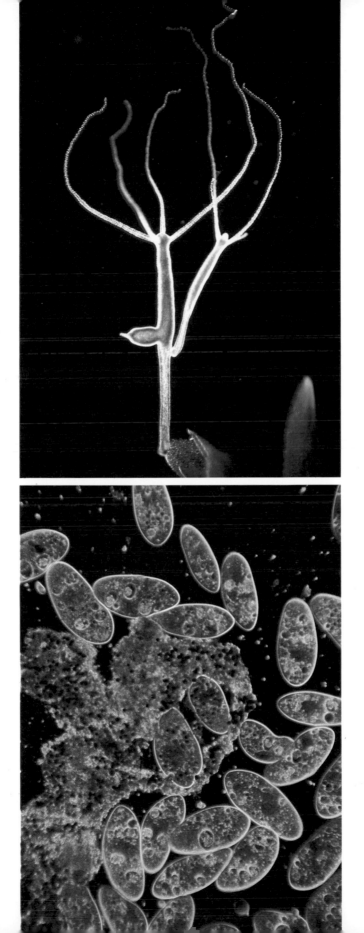

the naked eye appeared as the many-headed Hydra; a minute shrimp was christened *Cyclops* because it carried a single staring eye like the monster which Homer placed in Corsica. In drops of pond water this new breed of investigator discovered tiny, translucent creatures of all shapes and sizes, their bodies shimmering with urgently waving cilia. Polyzoans, rotifers, even bacteria came to light. One of Leeuwenhoek's students, Stephen Hamm, managed to observe sperms in a smear of semen, although he did not understand the significance of what he saw.

During the eighteenth century, the cataloguing of God's natural treasures begun by Ray and continued in a more ambitious way by Carl von Linné (Linnaeus) the Swedish botanist, was still far from complete. Gilbert White, another naturalist who had taken Holy Orders, was able to discover several new creatures even in the small country parish of Selborne during the second half of the century. He described the noctule bat and the harvest mouse, marvelling at the latter's ability to weave nests in standing corn or among thistles. To him, it was an elegant 'instance of the effort of instinct'. White was the first person to utilise behaviour in the form of song as a means of identifying bird species. Through voice patterns, he established the grasshopper warbler as a species; the cocks 'whispered' in the sallows. The sedge warbler's vocalisation was quite different from that of the reed warbler, with which it was often confused. White's real achievement was to sort out the leaf warblers or the 'willow wrens'. At the best of times, these little birds are most difficult to identify accurately. They are well camouflaged, and flit around in the depths of bushes or the tops of trees. Even the most accomplished bird-watcher armed with the best of modern binoculars, can sometimes glimpse only the wink of an eye or the flick of a wing before the bird melts away into the foliage. It is easy to suppose that the birds themselves have the same problem, until they start to sing. Gilbert White knew the notes of the two well. They were very different; one had a 'joyous easy laughing note', and the other a 'harsh loud chirp'. Today, the 'chirper' is called the chiff-chaff, and the almost identical bird with the joyful song, the willow warbler. A further variety gave White more trouble still because it haunted the beech woods that clung to the chalk escarpments around Selborne. He managed to 'collect' one, noting that it was larger and more yellow than other willow wrens. During the spring, this variety uttered 'sibilous grasshopper-like noises'. It was a common

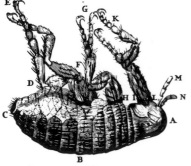

A flea engraved by Anton von Leeuwenhoek

enough kind of bird, but nevertheless new to science at the time, and became known as the wood wren or wood warbler.

Gilbert White's success as a naturalist stemmed from his intimate knowledge of a relatively small patch of countryside. He settled in Selborne as a curate in 1756. Selborne was not an accidental choice. His grandfather had been the vicar there, and he had himself been born in the rectory thirty-six years before. Then as now, the village was a picture of Arcadian bliss. It nestled in a fold of the chalk hills to the south-west of London, and was surrounded by beautiful beech woods and thyme-covered commons. Until his death in 1793, White lived the agreeable life of a well-to-do country parson. The statutory claims upon his time were far from onerous, so he was free to pursue his interest in natural history. As ever, it was considered to be an appropriate part-time activity for a rural cleric because it took him out into the fresh air where he could meet his parishioners, and appreciate the Glory of God through the perfection of His Creation. Exploring his parish, often on horseback, he collected accounts of country matters and ob-served wildlife. He also packed journals with data about the weather and rainfall.

White was an indefatigable correspondent, and many of his discoveries were recorded in letters he sent to a handful of friends and acquaintances. In those days this was one method with which to register fresh information. Eventually, he was persuaded by one of his more distinguished contacts, the Hon. Daines Barrington, to have some of his letters published. The result, *The Natural History and Antiquities of Selborne*, appeared in 1789. The book was referred to as 'the journey of Adam in Paradise'. And so it was, a testament of a modest man, content to deepen his understanding of one small corner of the world. To this day, it remains a treasured text for naturalists and a classic work of English literature.

In his letters he accurately described the habits of members of the swallow family, and those of the swift. In Selborne, swallows nested against stable rafters and in chimneys; house martins plastered their homes beneath the eaves of houses; sand martin colonies occurred in river banks; swifts screamed around the church tower. White noticed how young swallows emerged from the shafts of chimneys to sit on dead boughs while their parents fed them. He witnessed the way swallows sound the alarm when a hawk appears, thereby announcing to other birds the approach of danger. He also observed them skimming low over water to drink, and washing themselves on

the wing. In the autumn, the birds swarmed in large flocks around the church and its battlements as though 'preparing for their emigration, and consulting where to go'.

God was never far from his thoughts. Of nest building, in the swallow family, he believed that 'Providence had endowed different members of the same tribe with such varying architectural skills'. To him, it was a cause of wonderment. Nevertheless, he called for further research into the diet of swallows and martins to see if their different flying habits enabled the birds to intercept specific varieties of insect. This approach to behaviour, so different from previous attitudes, was one which was beginning to emerge among eighteenth-century naturalists.

Strangely enough, Gilbert White had nagging doubts about migration. Although his brother John wrote to him about the passage of birds past Gibraltar which he had personally witnessed, the parochial White clearly believed that the odd individual may have over-wintered in a cold, sluggish state. He wrote, 'House swallows have some strange attachment to water, independent of the matter of food; and though they may not retire into that element, yet they may conceal themselves in the banks of pools and rivers during the uncomfortable months of Winter.' The habit which swallows have in autumn of roosting in massive numbers in reed beds may have misled White and other naturalists. Also, swifts are occasionally overtaken by inclement weather and survive by turning torpid. However, White believed in the desirability of finding out the truth. One spring, he sent his gardeners thrashing through the bushes to flush out hibernating birds. None showed themselves!

With equal enthusiasm he turned to insect behaviour, and his account of the field cricket has hardly been improved upon since. Field crickets inhabited holes in the steep pasture behind Selborne where they basked and trilled in the warm afternoon sun between the middle of May and July. He even tried to transplant a colony onto the terrace in his garden, and kept one little creature in a paper cage from which it issued a merry, if irksome, song, manufactured by the brisk friction of one wing against the other. He found that they are solitary creatures, and that the males fight fiercely when they meet.

Gilbert White never asked too many questions about nature. To him, behaviour was a wonderful phenomenon to describe. In so doing, he managed to avoid the temptation of attributing human ambitions and emotions to animals. That was quite an achievement for a man who was, at heart, a poet.

A swallow drinking. The habit of drinking on the wing, and of roosting in reed beds, may have led people to believe that swallows passed the winter underwater

Birdsong as music in the
seventeenth century

In the intricately woven nest of
the harvest mouse, Gilbert
White perceived the wonders of
instinct

Meanwhile, there were people who were attempting to analyse in a more prosaic fashion what was going on in God's wonderful world. One of these was White's correspondent, the Hon. Daines Barrington. Barrington was a well-known judge, historian, journalist, and the creator of the prototype naturalist's journal. He was especially keen to discover how birds acquire their songs. A century before, John Ray had already suggested that 'old nightingales do teach their young their airs'. The experience of bird keepers who taught bullfinches to imitate folk tunes played on the flageolet also indicated strongly that learning plays a significant role in birdsong. Barrington placed nestling linnets into the nests of skylarks, woodlarks, and meadow pipits, and listened to the voices of the linnets when they fledged. Since they seemed to sing rather like their foster parents, he concluded that birds pick up songs by listening to their 'parents' whoever they may be. Unfortunately, his was such a sweeping generalisation that subsequent researchers found it easy to pick holes in his evidence. For instance, a cuckoo brought up in a reed warbler's nest still sings like a cuckoo!

What Barrington badly needed was a method of recording a linnet's song which would tell him at a glance how much of the fostered bird's voice was genuine linnet, and how much was picked up from the foster parents. He flirted with the idea of using musical notation. Composers had already incorporated avian melodies into their music, with mixed success; there was, for example, no general agreement about the interval between the two notes of a cuckoo's call! However, Barrington asked a harpsichord tuner to help him transcribe what he heard into patterns of crochets and quavers. Unfortunately, nothing came of the exercise, because much bird music defies the ear and their song does not conform to our own diatonic scale. The failure dashed all hopes of scientifically probing bird voices until the 1950s, when the technology of recording reached a relatively advanced state.

However, as the eighteenth century wore on, simple experiments were increasingly used to reveal truths about animal behaviour. One successful pioneer in this field was Lazzaro Spallanzani. Trained as a Jesuit, he was Professor of Natural Sciences at Pavia University, Italy, from 1769 until his death thirty years later. During his tenure, he managed to explode several medieval beliefs, including the theory of the spontaneous generation of life, which he refuted by laboratory tests. He discovered the crucial role of sperm in fertilisation,

using dogs as his experimental subjects. This paved the way towards the understanding of the biological importance of courtship.

Spallanzani also studied bats. It was thought that they employed supernatural powers to find their way in the dark. Spallanzani suspected that they possessed a mysterious sixth sense, and he issued a challenge to anyone who could discover its nature. Ludwig Jurine of Geneva responded, and reported the astonishing fact that bats with blocked ears could no longer detect obstacles in their flight paths. Spallanzani designed a set of experiments to test the accuracy of this observation. Catching some bats in the tower of Pavia cathedral, he blinded them to make sure that they could not use vision to orientate themselves. He also inserted a small brass tube into each of the bats' ears. In some, he left the tubes open-ended, but in the rest he obstructed them so that those individuals could neither see nor hear. When released, the blinded bats flew perfectly, as though nothing was amiss. By contrast, the blinded and deafened animals were utterly helpless, and fluttered into anything in their way. By this elegant if cruel experiment, Spallanzani confirmed the fact that bats required the use of their sense of hearing in order to navigate in the dark. (The ultra-sonic nature of the bat's signals, however, was not finally demonstrated until the twentieth century.)

Spallanzani was one of the last contributors to the Italian Renaissance. By his time, the centres of artistic and intellectual ferment had shifted to other parts of Europe. The process began during the early part of the seventeenth century when René Descartes began to speculate about the nature of life. His conclusions shaped the whole course of scientific thought for centuries. In France his disciples included influential leaders in biology such as Buffon, Cuvier and Lamarck. Descartes was loyal to the Church, and argued that the physical material of the Universe created by God behaved according to natural and definable laws which could be solved by scientific research. The method he suggested was to solve simple problems first, their solution leading inevitably to the elucidation of more complex ones. Descartes understood that progress could only be made by framing questions to which concrete answers could be found – an important qualification. Descartes also sought to explain the workings of the animal body as a system of mechanical operations unconnected to anything in the way of mind, consciousness or feeling. He was the ultimate 'mechanist', and could well have been the person who claimed to kick his dog periodically 'in order to hear the machinery squeak'. Descartes' philosophical theory embodied an explanation for the differences in behaviour and ability between mankind and animals. He asserted that the gulf was due to the presence of a reasoning human soul, which placed humanity on a superior plane to the rest of creation. This fundamental separation of man from other animals was a tenet which pervaded eighteenth and early nineteenth century thinking. It was certainly the view of Compte Georges Louis Leclerc de Buffon who dominated zoology during the latter part of the eighteenth century. He was superintendent of the Royal Zoo at Versailles, and wrote a monumental work in forty-four volumes, *Histoire Naturelle*, which purported to embrace all scientific knowledge. He clung to the belief that creatures were virtually mechanical contrivances, devoid of the ability to reason and reflect. He was, however, opposed in his extreme views by a remarkably well-educated gamekeeper who worked on the estates at Versailles and Marly, and who had time to compare and contrast the behaviour of wolves and red deer.

Like all gamekeepers, Charles Georges Leroy's worth was measured by the size of his employer's game bags, and by the daily addition of vermin to his gibbets. He was essentially a nature detective, able to read the signs in the grass, and to tell from the demeanour of the precious deer the extent to which they were, or were not, disturbed. Leroy also had to contend with the natural cunning of wolves whose predation deprived the aristocratic huntsmen of their sport. Finding nests of game birds; guarding warrens with their 'republic of rabbits'; pitting his wits against carrion crows, stoats, and birds of prey – all these imbued Leroy with a good practical understanding of the ways of animals. He published an account of his observations in 1764, in the form of a series of letters, as Gilbert White was to do. With typical French finesse, he wrote them under the *nom de plume* of 'the naturalist of Nuremburg' and addressed them to an anonymous lady who was possibly Madame d'Angiviller, Leroy's mistress.

Leroy was fascinated by the clear-cut difference between the behaviour patterns of predators and their prey, and speculated on how this affected their respective mental processes. He must have watched pairs of wolves hunting together, using all manner of ingenious tactics to intercept their quarry, returning later to their dens to nurture their families. The 'naturalist of Nuremburg' found wolves to have a very interesting 'moral

character'. Although living by slaughtering deer and devouring their torn and bleeding carcases, they rarely vented the 'cruel' side of their nature on each other. On the contrary, they often showed great affection towards their partners and cubs. Leroy argued that animals whose routine alternated between 'sleep and lonesome labours' could never be susceptible to tender feelings of compassion of the order that he observed in his wolves. Leroy attributed the apparent high degree of intelligence in wolves to the permanent association between males and females for hunting purposes, which 'gave birth to new ideas'. Here he anticipated the modern view that societies of the kind found among primates and social carnivores permit the ready transmission of acquired skills from one generation to another – culture.

By contrast, the stag's way of life, a relatively uneventful existence of calm and peace, militated against the acquisition of many 'new ideas'. 'He needs neither the ability to reflect nor to judge in the satisfaction of his wants. He is virtually surrounded by food, and soon learns where to find tender buds and shoots in the early spring, fresh juicy grass in summer, berries in autumn, and briars and shoots of furze when winter hardens the wood and withers the grass.' The bellowing of stags during the autumn rut was a sign of lust rather than love. Leroy noted that it involved no apparent sexual preference on the part of the stags; any hind would do to satisfy their craving for sex. They certainly exhibited no concern for their fawns. During the winter, the deer herded together only for mutual warmth, but afterwards separated, except for the fawns and hinds 'whom weakness and timidity kept together'. In forming temporary relationships involving only simple sensations, the mighty stag had no need to reflect or to acquire a vast store of knowledge; this would burden his mind to no purpose.

Leroy realised that the mere accumulation of isolated facts about an animal's habits was of little practical significance. He advocated that those who wished to understand behaviour should live with animals and observe their daily conduct. It should then be possible to assess which habits were 'due to the active influence of instincts and which were the result of intelligence'. He entreated naturalists to compile complete biographies of every kind of animal, setting a good standard with his own studies on red deer and the European wolf. He was also the first person to study animal behaviour in a comparative way, a practice that did not fully emerge until the beginning of the twentieth century.

Wolf cubs are treated with great affection by their mother. The French naturalist Charles Leroy supposed that this enriched their minds with 'new ideas' and helped to account for the wolf's intelligence

Leroy's interest in animal societies may have been heightened by the discontent among the French peasantry which eventually erupted in 1789 in the Revolution. Had Leroy not died of old age three years before, he might well have been guillotined for his connections with the aristocracy.

By the end of the eighteenth century, the boundaries of the known world were fast extending. With every new sea voyage and overland expedition, more of God's wonderful treasures were being revealed. For Europeans, the process started with the Portuguese and Spanish navigators like Bartolomeu Dias, Vasco da Gama and Christopher Colombus. By the time White and Leroy were observing wildlife, it had long been the practice to carry a naturalist on most voyages of exploration. This greatly facilitated the discovery and cataloguing of the new-found wealth of knowledge. For example, Georg Steller accompanied Vitus Bering, a Dane under orders from Peter the Great to discover whether Asia was joined to America. They came across a string of islands – the Aleutians – pounded by the Pacific swell. Here they found a giant sea cow which was named after Steller. These great marine mammals grazed upon the kelp which grew in those waters. Alas, within a few years Steller's sea cow became extinct. Steller also recorded a strange 'sea monkey' – perhaps the sea otter. Steller's eider, a beautiful duck, still lives in the neighbourhood of the Bering Sea. Joseph Banks and Daniel Solander, a pupil of Linnaeus, sailed with Captain James Cook into the Pacific in 1768. In Australia they saw odd creatures as large as sheep but jumping like jerboas; they were kangaroos and could outstrip *Endeavour*'s greyhounds. Shortly afterwards, the duck-billed platypus was discovered and those who first saw this strange compromise between bird and beast preferred not to believe their eyes. In 1775 Cook took possession in the name of George III of a bleak patch of Antarctica which he christened South Georgia. There he encountered the saffron-collared king penguin and a host of southern seals.

Books written by intrepid travellers were eagerly bought by people who thirsted for knowledge of far-away places, and who sought a little vicarious excitement to enliven the boring routine of their daily lives. Some authors became enraptured by their own fantasies and spun yarns of immense bravery amid overwhelming danger; they wrote of savage white bears which loped menacingly over northern ice fields, and ferocious apes in darkest Africa which fell from trees upon the unwary traveller. Tales of South America were especially popular

A stag bellowing during the autumn rut

CHARLES WATERTON

Charles Waterton (1782–1865)

during the early part of the nineteenth century. It seemed a particularly remote continent, one whose romantic appeal was enhanced by rumours of El Dorado and gold galore.

Charles Waterton's books were widely read. Waterton was one of England's great eccentrics. He was born into a family which claimed descent from the Sir Thomas More who was beheaded for opposing Henry VIII's marriage to Anne Boleyn. The Watertons had adhered to the Roman Catholic faith of their martyred ancestor, and had consequently suffered many indignities over the centuries from religious prejudice. By the time Charles was born in 1782, the worst excesses of victimisation had been curbed. Nevertheless, he was denied a seat in Parliament, or a commission in the armed forces, and was over-taxed to boot. He therefore understandably developed a profound and garrulous dislike for anything 'Protestant', including the reigning king. Waterton identified him with the brown or Hanoverian rat, which he loathed; it was the only creature upon which he waged unrelenting war.

From an early age, he had plenty of chances to become familiar with nature in the grounds of the family's Yorkshire mansion. As he grew, he also displayed an inclination to be boisterous and to play pranks. His behaviour was full of surprises. He clambered bare-footed and with ape-like agility to the tops of trees, a habit he retained all his life. In 1817, he celebrated a visit to Rome by scaling the angel surmounting Castel Sant-Angelo. Beds were not to his liking; he always slept on the floor, with his head resting on a beech-wood block. Knowing that cool water soothes sprains, he once attempted to cure a particularly bad one while he was in North America by holding his ankle in the thundering cascade of Niagara Falls!

Being very well read and religious, Waterton's attitude towards natural history was probably influenced by the Rev. William Paley, who had succeeded in raising nature study to new and dizzy heights of pious respectability. In 1802, he published his thesis on natural theology. The essence of his argument was set out in the front of his book in an analogy, 'The Watch on the Heath'. Were a country rambler to discover a beautifully-made watch on the ground, he would wonder how it got there, who had originally conceived its design, and who made it. Paley suggested that this kind of reasoning could be applied to the whole of the natural world. All animals and plants revealed signs of being lovingly and thoughtfully de-signed, incorporating clever features to ensure their survival. The evidence of design pointed to a designer – the Creator.

Two scenes from Charles Waterton's life. An experiment with rattle snakes (above), and Waterton riding a captured alligator in South America. From a biography of Waterton by Richard Hobson

The purpose of nature study was therefore to show that God existed. This was an argument that impressed a great many ordinary people who suddenly saw it as their duty to study nature, and to raid the countryside to make their own collections of flowers, ferns, and animals.

Two years after Paley had exhorted Christians to inspect the works of God, Waterton was in South America, managing the family's sugar plantations in Guiana. For the next twenty years, he divided his time between America and Europe, especially his Yorkshire estate, which he inherited in 1806, becoming the twenty-seventh Lord of Walton Hall – the Squire. Waterton was a pioneer among travelling naturalists, and had a mania for 'wandering' in South America. He found no gold but discovered a world teeming with objects calculated to increase 'our thanks and gratitude to God'. His first impression of the swampy jungle skirting the Demerara River was vividly recorded in his journal. He wrote of ravishingly coloured orchids, and humming-birds whose metallic tints surpassed those of the finest jewels. It was a feast for his eyes and ears. Flocks of scarlet parrots burst from the tree tops. Toucans rent the air with their yelps, and the rhythmic song of the white campanero bird sounded like the chimes of a convent bell. At sundown the bats flickered from their retreats, and nightjars – or goat suckers – hawked for moths in the light of his camp fire. As the inky darkness closed in, a chorus of piping and croaking frogs stunned his ears. Two hours before day break, Waterton heard the red howler monkeys start to growl and moan as though in dire distress. Of their vocal behaviour he wrote, 'You would suppose that half the wild beasts in the forest were collecting for the work of carnage. One of them alone is capable of making all these sounds, and the anatomist on an inspection of his trachea will be fully satisfied that this is the case. When you look at him, you will see a lump in his throat the size of a large egg; and if you advance cautiously and get under where he is sitting, you may have a capital opportunity of witnessing his wonderful powers of producing these dreadful and discordant sounds.' A male howler monkey does indeed produce one of the loudest noises in the animal world, a sound which penetrates a long way through the forest. Apart from accurate and charming accounts of behaviour, Waterton's books also contained stories of deadly poisons used by Indians – he discovered the source of curare – of thrilling encounters with monstrous snakes and of a rumbustious wrestling match with a giant alligator which he wanted to procure for preserving.

The raucous, far-carrying calls of the howler monkey help to ensure that family parties are evenly spaced throughout the rain forest

Waterton's enthusiasm for natural history and exploration knew no bounds. He had a scathing disregard for 'closet' naturalists, and like Leroy was firmly of the opinion that the best place to study animals was in their native haunts. That was a comparatively novel idea: the current zoological fashion was to shoot everything on sight and to study the specimens at leisure in museums. Waterton's sensible attitude is illustrated by his study of that strange creature, the sloth, which he came across on his jungle rambles.

Many people considered newly discovered species to be oddities, but not Waterton. 'Every species in the great family of animated nature is perfect in its own way, and most admirably adapted to the sphere of life in which an all ruling Providence has ordered it to move.' However, the sloth appeared to be an exception because it seemed 'forlorn and miserable, so ill put together, and so totally unfit to enjoy the blessings which have been so bountifully given to the rest of animated nature'. Waterton therefore decided to investigate the animal to see if nature had committed a blunder. However, he did not make the mistake of bringing the sloth into his home; he went out into the damp, sticky forest. He reported, 'The sloth is doomed to spend his whole life in the trees, and, what is more extraordinary, not upon the branches, but under them. He moves suspended from the branch, he rests suspended from it, and he sleeps suspended from it. To enable him to do this he must have a very different formation from that of any other known animal. Had he a tail, he would be at a loss to know what to do with it in this position: were he to draw it up between his legs it would interfere with them, and were he to let it down it would become the sport of the winds. Thus his deficiency of a tail is a benefit to him. His hair is thick and coarse at the extremity, and has the hue of moss which grows on the branches that it is difficult to make him out when he is at rest. Hence his seemingly bungled conformation is at once accounted for; it is fair to surmise that it just enjoys life as much as any other animal, and that its extraordinary formation and singular habits are but further proofs to engage us to admire the wonderful works of Omnipotence.'

Many people disbelieved Waterton's elegant account of the sloth and even doubted the existence of such a strange animal. Waterton responded with the declaration that sooner or later sloths would be exhibited in zoos and then everyone could see for themselves what he had observed in the South American jungle. He was right.

One of nature's mistakes? Charles Waterton thought that the three-toed sloth was beautifully adapted to its life in the trees

Waterton eventually settled permanently at Walton Hall, where he displayed a touching degree of benevolence towards his animals. He built his stables so that the horses could 'talk' to each other, and his sties were so arranged that the pigs could enjoy the sun. He developed his skill as a taxidermist, and continued to indulge his passionate interest in birds. A great proportion of his income was spent in managing his estate for the benefit of members of 'the feathered tribe'. In 1826 he completed a three-mile-long wall around his grounds, so establishing the first bird sanctuary in the British Isles. Foxes and other carnivorous mammals were trapped and released outside the perimeter. Waterton also had trouble with poachers who came after his pheasants. He encouraged these game birds for their decorative rather than their gustatory qualities. To outwit trespassers, he nailed large numbers of wooden decoys to trees in which pheasants roosted. In the gloom, the dummies were mistaken for real birds, and the poachers wasted a lot of shot! Waterton erected towers for roosting starlings and built 'homes' for owls. His lake he heavily stocked with wildfowl — the Squire liked nothing better than to while away the time drifting on a boat, watching the Canada geese and marvelling at the dazzling plumage of the kingfishers.

To Waterton belongs the credit for being the first ornithologist to realise the value of concealed hides or blinds for watching animal behaviour. He converted many hollow trees into places from which he could observe birds, sometimes in real close-up, at arm's length. For early Victorian days, these were advanced techniques in practical natural history. Some of his research was carried out in the spirit of true science, as when he teased apart the regurgitated pellets of owls in order to find out what they had been feeding on. He also counted the visits a pair of blue tits paid to their nest over the course of a single day while they were raising a brood of voracious nestlings.

And yet Waterton had his blind spots. He criticised with undisguised contempt the work of James Audubon, the North American artist and naturalist, and he had no time for the accounts of giant apes (gorillas) brought back from Africa by George Du Chaillu. He also heartily questioned the truth of Edward Jenner's accurate description of the way a day-old cuckoo heaves the eggs and nestlings of its foster parents out of the nest so that it can monopolise the food supply.

Waterton did not yield easily to fresh ideas. But people like Ray, White, Leroy and Waterton himself, the practitioners of the new natural history, had carefully accumulated a mass of accurate information. This now required to be re-ordered and re-interpreted by a different brand of expert, who would suggest a totally new approach to nature, something altogether less passive than the idea that everything in nature is an illustration of Divine Wisdom.

In October 1845, a carriage bearing a notable passenger drew up at Walton Hall. The guest was welcomed warmly, being spared the Squire's more outrageous forms of greeting. (He still enjoyed a good joke, and occasionally entertained his visitors by behaving like a mad dog, springing out from beneath a table, and trying to bite their ankles!) This time he behaved himself and treated his guest to an abstemious meal – Waterton never touched alcohol. They were joined by Waterton's sisters-in-law, and two Jesuit priests. The Squire was in good form, and entertained the company with well-worn stories of how he had valiantly ridden an alligator and tempted vampires to nibble his toes. Doubtless, he praised God for the fascinating world they were privileged to enjoy. The guest was of a more modest disposition. He too had travelled around South America as a ship's naturalist, and he had even voyaged to an isolated archipelago which barely registered on the maps of the time – the Galapagos Islands. He would have been too polite to voice his thoughts about the origin of species which that trip to those remote islands had generated. His name was Charles Darwin, and he was poised to present an enduring challenge to the faith of his host and those like him, who believed that animals and humanity were poles apart. The effect Darwin was to have upon scientific thinking, and upon the study of animal behaviour, was to be nothing short of revolutionary.

SEARCH FOR THE MIND

Charles Darwin (1809–1882)

In the industrial midlands of England, an oak tree stands beneath a perpetual pall of smoke. Once its bark was festooned with a flourishing growth of mosses and lichens against which a host of insect lodgers were superbly camouflaged. Now its trunk is bare and blackened with grime and soot leaving a beautifully spangled peppered moth naked and vulnerable in its day-time resting place. A foraging great tit spots the juicy insect and quickly devours it. However, the bird completely overlooks another peppered moth, rather unusual with its uniformly dark-brown wings. Because it blends into the background, the colouration of this 'sport' has greatly enhanced the moth's chances of remaining hidden. Since 1848, when the first melanic peppered moth was discovered near Manchester, the original variety has been preferentially eliminated by the birds which prey upon it from woods polluted by industrial fall-out, to be replaced by the melanic variety, better adapted to survive on darkened tree trunks.

Charles Darwin was the first person to grasp the fact that species had the capacity to change in this way. He also assumed that patterns of behaviour, as well as of anatomical design, were forged on the anvil of evolution, and postulated that man's own customs and habits had their origin in his animal past. Man's direct ancestors have long been extinct, but traces of his behavioural roots may still be recognised in the habits of his closest cousins.

This unprecedented train of thought led Darwin to the ape house of the London Zoo in about 1850, a decade before he achieved his considerable notoriety in Victorian society. He was an indefatigable and methodical researcher, who never

The snarl of rage of the male mandrill. Darwin saw in the facial expressions of primates the glimmerings of human-like thoughts and emotions

Facial expressions of a Celebes
'ape', as recorded by Darwin

lost an opportunity to jot down notes, record fresh facts, and carry out tests. So it was with his visit to the zoo. Not content simply to stare and marvel at what was then the most magnificent zoological collection in the world, he went armed with a note-book, pencil, and a mirror. He entered a cage housing a pair of young orang-utans and introduced them to the mirror. To onlookers, it must have seemed a crazy thing to do, but it had a serious purpose. Darwin was fascinated by facial expressions because he felt they might provide clues to the inner workings of the animal mind. Giving apes a mirror was a useful way to encourage them to pull faces and so to reveal what they were thinking and feeling. All creatures find mirrors puzzling because the reflections do not behave like normal companions; they mimic rather than respond! Darwin carefully recorded the reactions of the orangs. 'At first they gazed at their own images with surprise, approached close and protruded their lips towards the image, as if to kiss it. They next made all sorts of grimaces, placed their hands at different distances behind it, and looked behind it.' Obviously the young apes thought that they were seeing other individuals through the glass, not merely their own reflections.

Charles Darwin's curiosity about animal behaviour was not a passing fancy. He was the first of a new generation of naturalists who had a powerful desire to understand and explain whatever they saw. Darwin also believed that observations were useless unless they were connected to a theory of some kind. The theory which stimulated him to collect information was itself to solve what scientists of the day called 'the mystery of mysteries' – the origin of species.

Darwin was reluctant to submit his ideas to public scrutiny, but his hand was finally forced by Alfred Russel Wallace, a naturalist who had spent much of his time in the tropics, and who had independently formulated a similar theory of evolution. On 24 November, 1859, Darwin's book *On the Origin of Species* was published, and was an immediate success. The reason for his delay of nearly twenty years was abundantly clear. Renaissance scientists had already expelled God from the visible heavens: now Darwin proposed as the creator of species a pitiless, random process in place of the hand of Divinity. He spelt out with great verbosity his evidence that the environment promotes continuous development of living creatures. It was common knowledge that animals and plants produce far more offspring than could possibly survive. Darwin perceived the implication of this truth, namely that there must be a 'struggle for existence' between animals of the same species. He also knew that individuals of the same family vary among themselves – the peppered moth is an obvious example. This led him to the inescapable conclusion that only those offspring with variations that assisted survival – or the 'fittest' – would win through, breed, and so hand on their helpful characteristics to their descendants. Although a number of people before Darwin had suggested the possibility of evolution, all had failed to see how change could be brought about. Darwin's brilliant achievement was to provide a plausible theoretical framework for the process that drove evolution along. He christened it *natural selection*, or, to use Herbert Spencer's phrase, 'survival of the fittest'. It alone appeared to account for the phenomenon of 'descent with modification'.

Although Darwin was rather reluctant to discuss mankind, he assumed that humanity is part of the 'tree of life', sharing to a greater or lesser extent a common history with its fellow creatures. He certainly believed that there was mental continuity between man and animals, an outrageous notion that was hard for many devout and God-fearing Victorians to swallow. Many converts were happy to admit the possibility that evolution happend to 'brutes'; but they claimed that man, with his fine sense of morality and marvellous mind, was special.

The defeat of the over-confident Bishop of Oxford, Samuel Wilberforce, in the British Association debate with Thomas Huxley in Oxford on 30 June 1860 marked a watershed in the acceptance of evolution. Overnight, the study of animal behaviour changed from a frivolous pastime which might help to reveal the mind of God, to an activity worthy of scientists and philosophers for what it might reveal about the origins of our own customs and consciousness. Suddenly it was possible to analyse the relationship between the pout of an orang-utan and that of a sulky child!

Darwin's own study of behaviour was well under way when *On the Origin of Species* was appearing in the book shops. Naturally he was keen to discover 'how habits had been gradually acquired', and sought an evolutionary interpretation for man's own rich repertoire of gesture. He was of the opinion that expressions were merely 'emblems for the emotions', and he attempted to establish the 'states of mind' that were responsible for them. This was a formidable task; the contemporary view was that the whole matter of expressions was utterly inexplicable. Nevertheless, Darwin rose to the challenge and

Darwin noted the similarity between a pouting chimp and a disappointed child

Cat expressing aggression

Darwin used photographs to analyse facial expressions, as in this plate from *The Expression of the Emotions in Man and Animals*

pioneered several methods for analysing the subtle and fleeting movements that we call expression.

Darwin watched infants: 'they exhibit many emotions with extraordinary force'. Even his own children came under his scrutinising eye, especially his first son, William, who was born in 1839. His inquisitive father took delight in startling the lad with a rattle, duly noting that he appeared to smile and frown by instinct. Darwin was able to construct a schedule on which expressions and emotions developed. From the eighth day after birth, a slight puckering of the brow heralded a screaming fit; tears first glistened on the baby's cheeks when he was 139 days old. Another baby Darwin studied wept tears when 104 days old.

The developing art of photography was immensely useful to Darwin in teasing apart the differing roles of the separate muscles which power our facial expressions. He hoarded pictures of all kinds. His scrapbooks still exist, and among the well-worn pages are fading sepia-tinted snaps of bawling babies, demure little girls smiling sweetly (including one who inspired Lewis Carroll's *Alice in Wonderland*), guilty-looking ruffians, enraged men, raving lunatics, and coy women. Darwin even commissioned an actor who was a photographer to take pictures of himself in poses of defiance, anger, surprise, and so on. He returned repeatedly to a published collection of plates taken by a French surgeon, Dr Duchenne, who practised a rather fashionable if unpleasant technique which entailed applying a pair of electrodes to different parts of his patients' faces, recording the galvanically-induced grimaces for posterity.

Questionnaires returned from colonial administrators supplemented photographs in assisting Darwin with a comparative survey of the expressions and gestures of the races of mankind. He wished to find out whether similar actions express the same feelings the world over. If so, then a common underlying behavioural mechanism would be indicated, fixed by inheritance. Some gestures were obviously learned during early life, and differed in culturally separated groups of people; for instance, in northern India, agreement was signalled by an ordinary nod of the head, but in the southern regions the same message took the form of a sideways shake. Nevertheless, the similarities displayed by people of different cultures were striking. Darwin's conclusion still stands: that 'the same state of mind is expressed throughout the world with remarkable uniformity'. We all laugh in the same way, we all shake our fists in rage.

When faced with making sense of the daunting amount of information which had accumulated on the 'expression of passions' in animals, Darwin was strongly influenced by the fashion of the day. The laws of physics and chemistry dominated scientific thought; some researchers lived in hope of finding a Grand Theory that would account for everything, and doctors believed in the existence of a universal panacea that would cure all ills. In turn, Darwin attempted to create order out of chaos by identifying laws of behaviour which were generally true for both animals and ourselves.

In animal behaviour, Darwin saw the operation of an important evolutionary law leading to the advancement of life. 'Multiply, vary, let the strongest live and the weakest die.' It accounted for the fact that the males of so many species are more glamorous than their mates, often heavily decorated, or furnished with murderous weapons. He argued that the origin of these 'secondary sexual characters' lay in the urge to breed. Females tended to avoid mating with dull and dreary males, choosing instead, dashing and exciting partners as mates. Over the generations, this process of 'sexual selection' produced colourful and flamboyant males, especially among polygamous creatures such as pheasants and certain grouse. The 'season of love is also that of battle'. Darwin recognised that males were far more belligerent than females. Obeying the 'law of battle', the males seemed compelled to fight each other for their mate's favours. Zebra stallions stir up the dust on the African plains, chasing, biting, and kicking each other when there is a chance of taking a fresh mare in heat into their respective family groups; rival red deer stags roar at each other, and thrust and parry with their antlers when competing for hinds; even seemingly timid creatures like hares engage in desperate conflicts during the spring. It is little wonder that Darwin thought that male mammals won their females through the 'law of battle' rather than by the display of charms. His analysis revealed that mammals without cutting teeth, such as deer and antelope, carry special fighting armament on their heads.

Darwin thought much of animal aggression to be bluff rather than combat, the contestants attempting to appear large and ferocious by raising 'dermal appendages'. Indeed, the erection of bright patches of fur and feathers is widely used by animals to express their feelings. Cockatoos fan their crests jauntily when greeting one another; cob mute swans arch their wings in defence of their territories; a cornered cat bristles in terror; and a porcupine threatens enemies by erecting and rattling its

The 'law of battle'. Darwin recognised that males seem compelled to fight one another for females, and are often fearsomely well-equipped to do so

Opposite emotions produce quite different behaviour. A hostile dog (above) and a friendly one (below). From Darwin

menacingly sharp quills. When a dog approaches close to a hen and her chicks, she spreads her wings, ruffles her feathers, and, doubling her apparent size, rushes at the intruder.

Darwin's enduring contribution to the study of animal communication is his book *The Expression of the Emotions in Man and Animals*. Issued in November 1872, thirteen years after *On the Origin of Species*, it was delightfully illustrated with photographs and drawings from his collection. It was also the first major work of scholarship devoted to behaviour, and in it the foundations were laid for the two separate disciplines of experimental psychology and ethology, to which chapters five and six are devoted.

Darwin proposed three principles to account for expressions: *serviceable associated habits*; *antithesis*; and *direct action* of the nervous system. A few homely examples will clarify what he meant. A dog in a hostile mood walks stiff and tall; its tail is carried erect, the ears are pricked forward; its eyes stare fixedly at the opponent; and the lips retreat, revealing the fangs in a cruel snarl. The whole bearing anticipates the consequences of the dog's temper — to fight and bite — and is therefore adaptive or *'serviceable'*. When the dog is in an opposite frame of mind it may produce expressions which are the *antithesis* of the serviceable associated habit. A friendly, servile dog therefore cowers, its tail hanging loosely between its legs, and its lips drawn tightly over its teeth. Contrasting behaviour can also be observed in cats. A quarrelling cat assumes a crouched posture with the body and tail extended, ears pressed backwards and mouth open. From this position it can spit and strike out; the habit is 'serviceable'. An affectionate cat is a purring upright one, with its tail held vertically like a flagpole over its arched back, ears forward and teeth sheathed. It is the 'antithesis' of an angry animal. Through the course of evolution, ears turned back in anger for protection, are brought forward as a gesture of conciliation, and teeth exposed in fury to wound an enemy are hidden as a sign of friendship. Darwin even applied these principles to human behaviour. It is no accident that we bow and stoop before monarchs who sit on high thrones and wear full robes which exaggerate their size! Darwin also observed that certain states of mind were associated with actions which seemed neither 'serviceable' nor 'antithetic', such as the sweating or trembling induced by fear. He thought that these were useless byproducts of the workings of the nervous system — *'direct action'* — and mostly had no value as messages.

Today, no zoologist thinks of animal behaviour as complying with laws of the sort Darwin proposed. Nevertheless, in his behavioural principles Darwin pointed to an essential truth of visual communication. Angry animals do indeed flaunt their teeth and tusks, and brandish their beaks and antlers in front of their adversaries. However, the exercise of might and power is not the only way to influence a social companion. Special appeasement displays have been evolved which attempt to soothe an opponent and to. quench his anger. These often include movements which deliberately hide the 'weaponry'. A wolf therefore presents its vulnerable neck rather than its 'loaded' muzzle to another member of the pack in 'friendship'; many birds, in non-hostile frames of mind, turn their beaks upwards or away from their mates, thereby concealing their provocative battle visages as an overture of peace. Here lies the basis of Darwin's antithesis principle; signals which convey very different intentions, must, in the interests of clarity, be easily distinguished from each other.

Clear though Darwin's accounts of behaviour are, his style differs markedly from modern zoological writing. This stemmed from his method of analysing behaviour. When confronted with an animal in action, he described its movements in terms that presupposed that it had human feelings and emotions. Ants *despaired* after losing their home to the spade. Dogs took *pleasure* when doing what they considered to be their *duty*, such as when they carried baskets, picked up sticks, and retrieved game. *Shame* was a sentiment that he had seen expressed on the face of a dog which had done *wrong*. Bitches dearly *loved* their pups, and showed their *affection* to their mistresses. Terms like these would be deemed to mark his book as thoroughly unscientific. Nowadays they are excluded from scientific accounts on the grounds that it is unknown if, and undemonstrable that, glimmers of moral sensibility and sensation register in an animal's consciousness. Indeed, some zoologists even question whether animals are 'conscious' at all. However, Darwin was correct in arguing that the 'expression of emotions' was vital to the welfare of social species. Through facial, gestural, and vocal signals, animals attracted each other, repelled each other, stayed together, bred, and generally regulated their day-to-day relationships. He also found an overall similarity between the facial expressions of humans and those of our primate relatives, which led him to conclude that we share similar states of mind. In 1871, he wrote in his book *The Descent of Man*, 'The difference in mind between man and higher animals, great as it is, certainly is one of degree and not kind.' The prophet of evolution could hardly conclude otherwise!

Darwin's important and highly influential opinion meant that our fellow creatures were no longer regarded as unthinking slaves of inborn urges. Now that it could be admitted that animals were guided by more than instinct, there was a danger that their intelligence would be greatly exaggerated. Unfortunately for the study of animal behaviour, this is precisely what happened.

For the last two decades of his life, Darwin chose to become more or less confined to his home, Down House, in Kent. He was therefore unable to observe animals in person. However, he made up for his sedentary existence by becoming a prodigious correspondent, and by talking to naturalists who called on him. In 1871, he received one of the few overseas visitors to make the pilgrimage to Down. Lewis Henry Morgan, an American, had amassed a great deal of evidence to suggest that beavers were very intelligent creatures.

The countryside is full of animals which appear to exhibit quasi-human skills in the course of their daily lives. The beaver stands out from all rivals as one of nature's most accomplished engineers. Anyone who has observed a beaver meadow can be forgiven for thinking that these amphibious rodents work to an intelligently-conceived plan. The dam is the beaver's principal construction. Built from an interlocking mesh of sticks and mud with the tensile strength of reinforced concrete, a large dam can raise the upstream water level by as much as five or six metres. Somewhere in the pond which forms behind this barrier, the beaver locates its lodge. This too is a marvel of design. It is built rather like an Indian tepee, with an underwater entrance or two, an internal platform just above water level, and in the roof a ventilation 'chimney'. Surrounded by water, the beaver's home is practically unassailable, a castle protected by a moat. Radiating from the pond, the beavers maintain a network of canals about a metre wide, and these often run for hundreds of metres through rank vegetation to higher and drier ground. They use these waterways to reach the aspens, willows and birch trees upon which they feed, transporting some of the foliage back to their lodge. It was not surprising that such sights impressed Morgan. He was, among other things, an engineer, and so fully appreciated the beaver and its works.

Lewis Morgan was a remarkable man. His paternal ancestors

A beaver dam, with its lodge safely located in the middle of the pond

had emigrated from Bristol, England in 1648, and were among the founders of New London, Connecticut. Born in 1818, he was one of a family of thirteen children, and grew up near Lake Cayuga in New York State, where his father, Jedediah, served for a period as State Senator. Nearby, there was an Iroquois settlement and this kindled in Morgan a life-long interest in Indian culture. He eventually became the first North American social anthropologist, living among various tribes and recording their customs. As a lawyer, he championed their cause, fighting their merciless exploitation by the advancing tide of white settlers. In 1851, at Rochester, he married his devout cousin Mary, and became a stalwart Republican pillar of American society, and a prosperous business man. It was business that brought him to beavers.

Morgan helped to survey and finance a railroad to the area on the south shore of Lake Superior so rich in iron ore, for transporting the valuable mineral to regions where industry was expanding. The track had to be laid through forty miles of rugged wilderness, traversed by rivers and streams in which beavers abounded. Morgan was already fascinated by the subject of animal mentality, and believed that the 'Thinking Principle' was not the sole prerogative of man. In the swampy backwoods he had a heaven-sent opportunity to prove the point. Here lived vast numbers of creatures which displayed astonishing skills of engineering, architecture, and water management. Instead of spending his leisure time fishing for brook trout, Morgan set about making a methodical study of beaver culture. With the help of an Indian guide, and the Reverend Josiah Phelps, a keen amateur photographer, Morgan surveyed the network of channels, plotted the locations of the lodges, measured and photographed the dams, and counted the sharpened tree-stumps which marked the areas where the industrious animals had clear-felled the vegetation. In 1862, his unrestrained curiosity took him on an expedition up the Missouri River to the Rocky Mountains where he could compare the activities of beaver frequenting the steep valleys with those that lived in the flatter localities around Lake Superior.

Morgan's painstaking research was collated in the form of a book, *The American Beaver and His Works*. For the period, it was the most thorough study of a single species ever undertaken, rating as a milestone in the discovery of animal behaviour. Morgan's descriptions bear reading even today. His only error was caused by his conviction that beaver behaviour was the product of intellect rather than instinct. In the construction of the dam, Morgan saw intelligence in operation. 'The practice of beavers, while moving their short cuttings by water, of placing one end of each cutting against the throat and pushing it from behind, of carrying mud and stones under their throats, holding these there with their paws, and of packing mud upon their lodges by a stroke of the tail are intelligent acts performed sensibly and rationally.' Every now and again, beavers pause in their labour 'evidently to see whether it is right, and whether anything else is needed'. In this, 'he shows himself capable of holding his thoughts before his beaver mind; in other words he is conscious of his own mental processes'. Of the beaver's canal system, Morgan stated that 'to conceive and execute such a design is one of the highest acts of intelligence'. And in the tree-logging operations in which these animals tend to indulge during the autumn, he saw the application of his 'Thinking Principle'. 'A beaver seeing a birch tree full of spreading branches which to his longing eyes seemed quite desirable, may be supposed to say to himself, "If I cut this tree through with my teeth, it may fall and I can secure its limbs for my winter subsistence." But it is necessary that he should carry his thinking beyond that stage and ascertain whether it is sufficiently near to his pond or to some canal to enable him to transport the limbs to the vicinity of the lodge.' Morgan watched, and wondered at, the beavers' labour, busily dismantling trees and storing their terminal twigs and branches in sub-aquatic heaps to provide nourishment in the winter when the surface of the water was iced over. He also took this behaviour to be proof of a refined mind. 'It shows a forecast of the future. To satisfy present hunger is a simple act of intelligence; but to anticipate distant wants and provide for them is a much higher act of knowledge.'

From his experience of beavers, Morgan thought that they possessed a 'mental principle' which manifested qualities similar to those displayed by the human mind. They appeared to be conscious of problems, which they solved in a sensible fashion through the exercise of reason. Charles Darwin was well aware of Morgan's interpretation. When the Yankee businessman called on him, he politely expressed the view that Morgan had been deceived by the power of 'instinct' to produce behaviour which, at first sight, looked amazingly resourceful.

Modern scientists have revealed that Darwin's intuition was correct. Concentrating on events which trigger reactions, they have discovered that beavers are bothered by the sound of running water. In the laboratory, they try to escape from it. In the wild, these animals attempt to stop the noise by placing dams where water flows fast. Nothing brings a beaver to its dam more quickly than the chuckle produced by a leak. Beavers can even be tricked into plastering mud and sticks over a loudspeaker from which issues the sound of trickling water! The beaver therefore behaves, at least in part, according to a fairly rigid, inborn instruction: where water is noisy, dam it!

A century or more ago, people were not privy to the secrets revealed so recently by experimental biologists. Darwin's theory sent most of his disciples scurrying to museums to examine for themselves the anatomical evidence for evolution. Field work of the calibre of Morgan's became unfashionable, a trend which greatly slowed the advance towards an understanding of animal behaviour. Nevertheless, his descriptions and interpretations were seized upon by those who were excited by the notion of mental affinity between animals and man. They also collected anecdotes about talking parrots, tool-using monkeys, and door-opening cats and dogs, which purported to demonstrate that our fellow creatures are capable of flashes of insight, and at least the glimmerings of a consciousness human in scope. Rather than watch animals, W. Lauder Lindsay, a London physician, trained himself to recognise what was seemingly true in eye-witness accounts of behaviour. He placed a great deal of faith in stories related to him by schoolgirls and farmers' wives, who he considered to be especially reliable! Unfortunately this sort of approach did not succeed in illuminating the roots of our emotions and the evolution of our mind, but too freely endowed animals with an intelligence little short of our own. By far the most influential of these 'purveyors of anecdotes' was George Romanes, who, through his research, developed a very close relationship with Charles Darwin.

Although born in Canada in 1848, George Romanes was brought up to enjoy the life of an English gentleman. His parents were wealthy, and he spent his childhood in London, Heidelberg, and Scotland. In the Highlands he developed a taste for hunting. Until he fell fatally ill in 1893, he was to be found every year on the Scottish grouse moors. With the help of private tuition, he entered Gonville and Caius College, Cambridge in 1867. Like Darwin, Romanes considered a career in the Church, but changed his mind when he won a scholarship in Natural Sciences.

After graduating, Romanes cultivated an interest in physi-

A beaver in one of its many canals. Lewis Morgan thought that these waterways were the ultimate achievement of beaver intelligence

ology, then the fashionable subject among young zoologists. He turned to jellyfish – later to starfish and sea urchins as well – determined to discover whether they possessed a nervous system of the kind found in more advanced creatures. He embarked on his research at University College, London, and pursued it during the summer in a little seaside cottage at Dunskaith in Scotland. Here he applied himself with zest, recapturing the happiness of his childhood, rambling over the weed-strewn rocks, peering into a thousand rock pools and watching the 'crystal globes' pulsating with life and gleaming with all the colours of the rainbow.

After a letter of his on an obscure evolutionary topic was published by the scientific journal *Nature*, Romanes was thrilled to receive an invitation from Darwin to visit Down House. They met in 1874. Darwin greeted him with great warmth, exclaiming 'How glad I am that you are so young'. He saw in the youthful Romanes, a zoological successor who could carry on where he had left off. At sixty-five, Darwin professed to feeling increasingly worn out and to lack the mental stamina to grapple with fresh ideas. Romanes sparkled with enthusiasm, had a good brain, and had youth on his side. The meeting marked the beginning of an unbroken friendship. Judging by the letters which passed between them, Romanes had a respect for Darwin amounting to reverence. Darwin, in turn, took a fatherly interest in the young man's work.

Spurred on by advice and small financial contributions from Darwin towards the cost of his laboratory equipment, Romanes sliced and dissected his way through countless jellyfish, electrically stimulating portions of them, and seeking clues to their anatomy from the slivers of tissue viewed through his microscope. He identified a network of nerves, and discovered how the pulsing swimming behaviour was regulated. It was meticulous research, and a major contribution to zoological knowledge.

In 1878, Romanes prepared for a lecture in Belfast on 'animal intelligence'. This led him to consider the simple behaviour of a jellyfish, representing one of the earliest forms of life on earth, and how the processes of evolution could have developed it ultimately into the complex manifestations of the human mind. Darwin encouraged Romanes in his task and even gave him some of his own notes on behaviour, together with the manuscript of a chapter on instinct which he had written for *On the Origin of Species*, but had omitted on the grounds of length. Romanes eventually included it in one of his own

Jellyfish swim by pulsating movements of their bells. George Romanes investigated the nerves responsible for initiating the swimming behaviour. Above, one of his illustrations

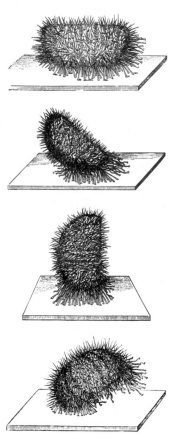

The simple behaviour of a sea urchin, which rights itself if turned upside down. After Romanes

A cruel cat? A cat gives the impression that it possesses a sense of cruelty because of the way it toys with its prey. But playing with its prey helps to make a cat more skilled as a hunter

books, *Mental Evolution in Animals*, published in 1884, two years after his mentor's death. Romanes immersed himself in his project, undertaking a vast and comprehensive review of the whole of the animal kingdom in order to provide a trustworthy measure of the level of intelligence attained by different kinds of creatures. In so doing, he hoped to trace 'the genesis of the mind', and to formulate a theory of mental evolution that would stand alongside and support the 'Darwinian hypothesis'. It was an awesome task.

If Romanes had investigated the behaviour of higher animals with the same scientific propriety with which he had studied jellyfish, he might have made great progress. However, when studying animal consciousness, he virtually rejected the experimental approach. The few tests he carried out were not an unqualified success. For example, he became interested in the homing ability of animals. He decided to experiment with cats and collected a carriage-load of pet animals from houses in Wimbledon. He then released them in the centre of the common, noting carefully the direction in which each animal set off. All he succeeded in creating was a chaos of disoriented cats! Reporting the results to Darwin, Romanes stated that cats were 'exceedingly stupid'. Everyone knew that these animals were perfectly capable of finding their way home; the pursuit of experimental proof was therefore pointless! This disaster confirmed him in his belief that tests could yield misleading results. He concluded that it was best simply to infer what was going on by 'watching mental phenomena'. Unfortunately, Romanes left it mostly to other people actually to watch animals, and allowed them to judge for themselves what was going on.

To verify claims about the intelligence of primates, Romanes borrowed a small South American monkey from the London Zoo and gave it to his sister to observe for a few months. She kept a diary of its general habits, and of the havoc that it created with the accoutrements of civilised Victorian living. When feeling especially mischievous, it ripped and smashed anything it could lay its hands on. When shown how to break open a walnut with a hammer, the animal quickly mastered the art. It also rapidly learned to unscrew the handle of a hearth brush, and to assemble it again. Shown a mirror, it clearly took its reflection to be a real monkey. When Romanes' sister pretended to feed a toy monkey, the zoo animal became very jealous and threw a tantrum.

The diary was included in Romanes' book, *Animal Intelli-*

Siamese fighting fish. George Romanes thought that their fighting was due to sexual 'jealousy', and that this emotion formed the basis of Darwin's 'law of battle'

gence. Published in 1882, it was enormously successful. It was a compendium of stories culled from respectable members of Victorian society: army personnel, explorers, naturalists (including Mr Darwin), as well as pet owners, zoo keepers and farmers. The accounts were arranged in order, starting with those about single-celled animals, and progressing to those about primates. For example, Romanes examined the theory that spiders were fond of music; he marshalled the anecdotal evidence that parrots knew what they were talking about, and experienced a sense of pride after they uttered words of wisdom. Fish were reported to experience jealousy. Romanes concluded that this emotion accounted for Darwin's 'law of battle'. Birds were said to feel affection and sympathy, and this contention was supported by stories of rooks failing to leave a dead companion after it had been shot, instead loitering in the vicinity, uttering distressful cries. These birds also displayed a sense of justice, which they manifested in the form of rook parliaments, in which they gathered around an errant flock mate and pecked it to death. Birds, it seemed, could also harbour grudges. Romanes documented the case of a parrot which was annoyed by a cat. The offended bird retaliated by seductively calling its feline adversary into the kitchen whereupon it seized a basin of milk and dropped it with great accuracy upon the poor beast. As the cat beat a hasty retreat, the parrot cackled with satisfaction. Despite his own experience with homing cats, Romanes collected evidence that cats were not entirely without intelligence. In the 'spontaneous' opening of latches, the ringing of bells, and the striking of knockers, cats demonstrated a creditable understanding of mechanical appliances. Dogs enjoyed a good joke; monkeys were curious and imitative.

With the help of anecdotal information, Romanes was able to construct a complicated process of mental evolution. He believed that emotions had their origin in the growing structure of the 'mind'. As nervous equipment became more complicated, animals became increasingly perceptive and capable of experiencing an ever-widening range of feelings. He even went as far as to name the emotions in order of their historical appearance. For example, surprise and fear were the first to register in animals low down on the evolutionary scale. More advanced creatures felt sexual affection, pride, grief, and hate. A 'sense of the ludicrous' was one of the last emotions to materialise. Furthermore, Romanes suggested the idea of a series of 'psychological states' to which different kinds of animals aspired. For example, spiders and insects which cared for their offspring were capable of experiencing similar emotions because of the social feelings which, he believed, parental behaviour aroused. Most molluscs were emotionally primitive, but cuttle-fish, squid, and octopuses were placed alongside the reptiles because of their ability to recognise people. Birds achieved a level which enabled them to experience an aesthetic love of ornament, as illustrated by the bower birds which decorated their mating arenas with flowers and feathers. Cats were in the category distinguished by the appearance of a sense of cruelty, whereas dogs and apes were placed in the highest state, as possessing a sense of shame, remorse, and a capacity for deceit.

Romanes' table of emotions was essentially a chart of the evolution of the mind from its first protoplasmic beginnings, the steps of which are retraced in man during the process of growing up. He thought that a newly-born child had the mental stature of a jellyfish or sea urchin, both of which, he believed, showed slight signs of self-awareness. A three-week-old baby appeared to be like a caterpillar, totally preoccupied with one activity: feeding. Romanes compared a human child at four months with a reptile, because both were capable of identifying individuals. By the age of fifteen months, the child already exceeded the ape in mental adroitness and in manual dexterity. Thereafter, the infant streaked ahead in the development of its mind.

Unfortunately, Romanes' method was faulty. He was altogether too gullible in accepting at face value interpretations of behaviour which credited animals with unwarranted degrees of intelligence — a mistake which many people still make! Nevertheless, his conclusions were much quoted by Victorian intellectuals who were constantly debating the nature of the thoughts and feelings experienced by animals. Yet, by falling so conspicuously into the trap of 'humanising' animal habits, he acted as an important catalyst for others more scientifically critical than he.

One of the most influential of these was his friend, Conway Lloyd Morgan, Professor of Zoology and Geology, later Vice-Chancellor of Bristol University. Tired of the 'anecdotage', he joined the debate in 1884. Observing the manner in which a terrier learned to open a gate, he discovered that its 'clever-ness' did not arise from an understanding of locks and levers, but originated in accidental movements as the creature wildly pawed the gate in its attempt to get through. With repetition, the useful actions became implanted, the ineffective ones were

A natural artist, the satin bower bird decorates its bower with an amazing variety of blue objects. This behaviour led George Romanes to suggest that birds had an 'aesthetic love of ornament'

abandoned. Finally, the dog performed perfectly, giving an erroneous impression that it had an intelligent appreciation of how to open gates! From his study of animals, Lloyd Morgan developed a precept which insisted that naturalists and zoologists opt for the simplest explanation of behaviour consistent with the evidence, avoiding any assumption concerning the involvement of higher mental faculties. Lloyd Morgan's influence was important in establishing a trend among scientists towards ignoring that which could not be proved, and in encouraging people to offer explanations of behaviour which left out of account an animal's supposed feelings and thought processes.

Romanes and his view of animal behaviour continue to be regarded sympathetically by pet owners, many of whom see their charges as 'people dressed in animal clothing'. However, in the long run, Romanes was scientifically discredited. Oddly enough, he was aware of, but chose to ignore, the conclusions of two people whose view of animal intellect differed from his own. In France and Corsica, Jean Henri Fabre was engaged in demonstrating that insects were less 'intelligent' than their remarkable behaviour might suggest at first sight. He constantly witnessed acts of remarkable skill: hunting wasps delivered their paralysing stings to the nerve centres of heavily armoured weevils with surgical finesse; mason wasps built their nests from mud and plant fibres with great architectural expertise. But despite their mastery of technique and technology, Fabre believed that his insects were driven by mindless instincts. He elegantly proved the point for pine processionary caterpillars. They behaved rather like sheep; where the first one went, others followed in a regular string, head to tail. This was how they moved from branch to branch in their constant quest for foliage. Fabre collected some, and carefully arranged them in a single file around the neck of a vase one metre in circumference. The caterpillars set off in a circular track – a journey with no end. They continued to lay down their silken trails for seven days and seven nights without a break! Nor could Fabre find signs of intelligent awareness in the behaviour of the mud-dauber wasp, which slaved away for days neatly camouflaging the outside of its clay nest with bits of moss and twigs despite the fact that it had chosen to build on a brilliant white wall!

Processionary caterpillars normally move from branch to branch. The French naturalist Jean Henri Fabre made them follow one another's silken trails, round and round in a circle, for seven days

George Romanes likewise ignored the research of Douglas Alexander Spalding, an ex-slater from Scotland, another who corresponded with Charles Darwin. He was a philosopher, and had been led to investigate behaviour because of Darwinism

and its implication of mental continuity between animals and mankind. Spalding was among those who believed that many skills in both animals and humans were 'born with us', and required neither a learning process nor the application of rational thought. It was a controversial subject. While argument raged among armchair contestants, Spalding decided to resolve some simple issues by experiment. In doing so, he became the first person to analyse animal behaviour in a scientific manner. It led him to demonstrate that behaviour previously thought of as being 'clever' was built into the mind from birth.

Comparatively little is known about Spalding. At twenty-two, he was fortunate enough to attend lectures in literature and philosophy at Aberdeen University without payment. Afterwards he travelled to London, where teaching brought him a modest income. This enabled him to study law, and eventually he was called to the Bar. Sadly, he had the misfortune to contract tuberculosis, and he was dogged by ill-health until his death at the relatively early age of thirty-seven. While in London, he struck up a friendship with the eminent liberal philosopher, John Stuart Mill, who felt passionately that ex perimentation was a powerful aid to scientific discovery. Through one of Mill's acquaintances, Spalding was introduced to the Amberley family and their circle of radical companions. Lord and Lady Amberley were notorious political and religious 'firebrands'. Impressed by Spalding's intellect and atheistic outlook, they engaged him as a tutor for their children. Violent objections were raised by Lord Amberley's father, Lord Russell, a former Prime Minister. However, Spalding's ability to handle Frank, the least manageable of the sons, finally won his grudging approval.

In 1873, Spalding moved, lock, stock and barrel, into the Amberley's home, Ravenscroft, in the Welsh border country. Apart from instructing Frank and eventually Bertrand – later to become famous as Bertrand Russell, the illustrious mathematician and philosopher – Spalding brought his own brand of chaos to the gentility of the mansion. He kept finches and bees in his small gabled bedroom, his chickens roosted in the drawing-room, and tame rabbits wreaked havoc on the lawns. Luckily, keeping animals in the interest of science was accorded the whole-hearted approval of the Amberleys, especially Lady Amberley, who enthusiastically embraced the task of assisting Spalding with his detective work.

Spalding wished to find out whether birds needed 'to rehearse

An expert architect, the mud dauber wasp builds its nest without having been taught to do so

their wings' in order to fly properly. He chose to experiment with swallows, shutting up five unfledged chicks in a small box not much larger than the nest from which they were taken. The box was hung on a wall close to the original nest site, and had a wire front so that the parents could still feed the imprisoned brood. In their cramped quarters, the young swallows had no chance to unfurl and exercise their wings. At the normal fledging time, Spalding and the Amberleys released them. Each bird rose effortlessly and flew with little trouble. After a few minutes in the air, the young swallows were sweeping around the trees, apparently with all the expertise of their parents. Spalding correctly concluded that flying is a skill which develops with the maturing nervous system, its appearance quite unconnected with experience.

Most of Spalding's observations were made on the young of domesticated species. He noted that young chicks were greatly alarmed by a tame hawk although they had never seen one before. From the moment it hatches, a turkey is able to stalk flies, and to catch them by means of sudden dashes. Spalding referred to this behaviour as a 'masterpiece of turkey cleverness', and he observed that it seemed to be perfected by practice. Chicks, on the other hand, never learned the technique although Spalding gave them plenty of opportunity to watch young turkeys demonstrate the art of fly-catching. He drew attention to the newly-hatched chick's 'instinct to follow' any moving object as soon as it was able to walk. But it had no greater inclination to attach itself to a hen than to a duck or a man. Much later, the German zoologist Oskar Heinroth re-discovered this phenomenon and called it 'imprinting' (*Prägung*).

Spalding was at his best when devising experiments to find out what abilities chicks 'are born with'. He wished to rule out the possibility of his birds learning from their mothers, so he artificially incubated the eggs on canvas cradles slung above pans of steaming water. When an egg cracked, he removed pieces of shell, and, drawing out the neck of the chick with its eyes still shut, he slipped a little hood over the head, fixing it by means of elastic threads. It was an operation that required the utmost delicacy. He kept them blindfolded for up to three days – an experience that did them no harm.

The circumstances in which 'these little victims of human curiosity' were first allowed to see were carefully arranged. Twenty were individually released in the centre of the kitchen table on which flies were placed, some alive, some dead. This

A young augur buzzard stretches its wings in readiness to fly. Birds do not have to learn to fly, but do so instinctively when the time is ripe

was intended to test their ability to judge distance when pecking at food. When the hoods were removed, the chicks were at first stunned by the glare, but soon keenly followed the moving flies with the precision of older fowls. Their first pecks at food were usually delivered accurately. The chicks did not attempt to seize objects beyond their reach 'as babies are said to grasp at the moon'. Darwin thought that a mother hen stimulated her offspring to peck at food by calling to them. Although she certainly does so, the day-old chicks clearly need no help.

Spalding also tested whether chicks have an innate capacity to recognise a mother hen's voice. One by one, he took chicks which had hatched in a darkened bag and placed them three paces from a coop in which was concealed a hen and her brood. Although the hatchlings had neither seen nor heard a chicken before, the chicks without exception set off for the box in answer to her clucking. Even hatchlings still wearing their hoods behaved in a similar manner, although they were slightly hampered by their blindfolds.

Pigs were given the same experimental treatment. From watching them in the barnyard, Spalding had a hunch that they also had their 'wits' about them at birth. He sat up one night waiting for a sow to farrow. The first piglet he placed in a sack and removed it from its mother's earshot. Seven hours later he released it outside the sty. 'The piglet soon recognised the low grunting of its mother and went without pause into the pig house, and was at once like the other piglets in its behaviour.'

Further experiments called for the temporary deafening of chicks by sealing their ears with several folds of gum paper before they finally emerged from the eggs. They were completely indifferent to the sight of the hen. After three days, when their hearing was restored, they ran straight to her, showing that it was the call, not the sight, to which they automatically responded. It confirmed him in his opinion that 'creatures bring with them into the world a good deal of cleverness'.

In the summer of 1874, Spalding's pioneering research programme ran into trouble when diphtheria struck the Amberley household. Frank recovered, but his sister Rachel and Lady Amberley succumbed. Spalding wrote to Darwin of his crushing sense of loss, adding dejectedly that his drive and enthusiasm for experimenting had evaporated. Shortly afterwards, the epileptic Lord Amberley died, leaving Spalding and another

Clever dog opening a door? In fact, such acts of apparent intelligence may be learned by trial and error

friend of the family as guardians to Frank and Bertrand, in the hope that they would be protected from 'the evils of a religious upbringing'. Lord Russell, the children's grandfather, successfully contested the will, and Spalding was left with virtually nothing. Tuberculosis was increasing its hold and he died in Dunkirk in 1877.

Such was the Russell antipathy towards him, that nearly all of the evidence of Spalding's life with the Amberleys was eradicated from the family records. There was one scandal that the Russells may have wished to cover up. The unorthodox Amberleys agreed that the ailing Spalding should not be allowed to marry and have children. Nevertheless, it was unfair to expect him to remain celibate. Lady Amberley, with the connivance of her husband, therefore allowed the family tutor 'to live with her'. The revelation of this shocking state of affairs was perhaps one of the reasons for Spalding's ostracism.

He may also have fallen into scientific obscurity after his death because he was enmeshed in the growing class war. Victorian England was nothing if not class conscious. Spalding's roots were buried deeply in the working class, and he was what today would be recognised as a Socialist. Furthermore, he was untrained as a scientist. He was therefore shunned by the great scientific institutions like the Royal Society, so he had to report the results of his neat experiments in *Nature*, and *Macmillan's Magazine* – a liberal, philosophical journal. Had the patronage of the Amberleys continued, he might well have been credited with founding the experimental science of animal behaviour.

Some of Spalding's contemporaries conjectured that the 'mind' and its manifestation, 'consciousness', spurred the body into action. Spalding was convinced by his chicks and piglets that behaviour, whether of man or beast, was in a sense automatic. He believed that even so-called 'voluntary' acts were produced by automatic, inborn nervous processes. Both animals and man were therefore 'conscious automatons', never free from whatever laws of nature controlled them. He might have been reinforced in his conclusion by familiar examples of 'mindless' behaviour. At night, insects were drawn compulsively to the glare of lamps. Moths appeared hell-bent on suicide, striking their wings and furry bodies against the glass chimneys of oil lights. If the flame was naked, the moths fluttered to their doom, flying around in ever-decreasing circles until they became singed to death. Then as now, birds migrating at night occasionally became trapped in the powerful beams of lighthouses, battering themselves to death. They seemed no more able to head off into the darkness and safety than moths could avoid the flicker of a candle flame. It was a German scientist, Jacques Loeb, who eventually provided an explanation for such puzzling phenomena, and eradicated the last traces of conscious mind from interpretations of animal behaviour.

Born in 1859, Loeb grew up during a period when Germany was the centre for medical and psychological sciences. He was an extraordinarily intelligent young man, fascinated by such philosophical questions as whether there was 'free will'. In 1884, at the age of twenty-five, he was awarded a medical degree, thereafter consecrating his life to science. With him, faith in the mechanism became a guiding force, and through research, he was determined to reduce everything, including noble acts of self-sacrifice in man, to mechanisms. The reputation of Romanes was still running high, but Loeb rejected the idea that animals possessed human-like minds and motives. However, he was at a loss to know how to prove it until he became an assistant to the professor of physiology at the University of Wurzburg. There, he was befriended by the eminent botanist, Julius von Sachs. Von Sachs had elaborated upon Charles Darwin's last major thesis, *The Power of Movement in Plants*, in which Darwin had likened the behaviour of moles and earthworms to the downward movement of roots in response to gravity. Darwin had even sent his son, Francis, to study under von Sachs. The German professor demonstrated that various plant organs oriented themselves towards or away from gravity, moisture, chemicals, solid objects, and light. These simple non-voluntary actions he christened *tropisms* or 'forced movements'. When Loeb met him, von Sachs was at the height of his career, but on the verge of a nervous breakdown, and heavily addicted to morphine. In the summer of 1887, they walked regularly in the park, where Loeb learned of the research which proved that, in their response to light, plants behaved like simple photochemical machines, bending this way or that depending upon the direction of the illumination. The thought occurred to him that perhaps there might be a correspondence between the behaviour of animals and plants. If only he could demonstrate examples of animal behaviour mindlessly performed in response to an external influence, like light.

Loeb lost no time in visiting the bowels of the local zoological museum to search for specimens which resembled plants. He discovered tube worms, and quickly procured some live ones

Jacques Loeb's experiment to measure the relative attractive power of lights of different intensity on barnacle larvae

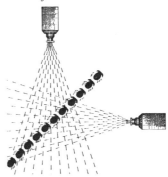

which he 'planted' in a sea-water tank. When deprived of the tubes in which they resided, the worms slowly built new ones angled towards the light, whichever direction Loeb placed it. They behaved just like von Sach's seedlings which always canted towards the sun. Sensitivity to light appeared to be a general phenomenon. The caterpillars of the moth *Porthesia* climbed to the tips of branches where they gorged themselves on the opening buds. Loeb conducted some simple tests which proved that they were 'slaves to the light'. It attracted them compulsively, so that they crawled towards the bright sky until they could go no further. Such was the pull of light that they would even starve rather than turn their heads into the shadow if food was placed behind them. The tube worms, caterpillars, and night-flying moths which gather around lamps, were said to be positively phototropic. Other kinds of creatures, such as blow-fly maggots and king crabs, were repelled by light, and did their utmost to escape from it.

In Naples, and later America, Loeb set about analysing the phenomenon with the precision and methods of a physicist. His tropism theory predicted that, when faced simultaneously with two lights coming from different directions, a positively phototropic animal should move on a course such that its eyes would receive equal amounts of illumination. If one of the lamps was reduced in intensity, then the heading should change to compensate for the unequal pull. His hunch was borne out by tests with the planktonic larvae of barnacles. When released in the beams of two lights, one of which could be dimmed in stages, they swam at an angle which could be predicted by a precise mathematical law. There was no evidence of free will, only of forced movements of a robot-like quality.

Loeb's research revealed all manner of tropisms. Shrimps aligned themselves to electric currents; delicate hydroids oriented to the tug of gravity; lusty male moths were lured compulsively to females by a chemical released from glands in their tails. His theory provided a basis for understanding these simple orientation movements of lower animals. But, in the flush of enthusiasm for tropisms, Loeb tried to apply it even to human intelligence, maintaining that ultimately we are all biological machines, compelled to react to physical conditions like corn waving in the wind. He saw little scope for 'voluntary behaviour'. To him, the mind and its manifestations were an irrelevance. As such, he was the antithesis of Romanes and his feeling, thinking, conscious animals.

Loeb, like many scientists, pushed his ideas too far. They

Tube worms behave like plants, as 'slaves to the light', and angle their tubes towards its source

Chinese silk moths mating. Lured by her sexual scent, a male moth compulsively homes in on the female

were too simple to explain the whole spectrum of behaviour. And yet, the fact that he sought common explanations for both human and animal behaviour was a legacy from Darwin, who had proposed the unthinkable – that animals and man have a common heritage and share a great many characteristics. Although Darwin instigated an explosion of interest in animal manners and habits, no simple explanations endured. It was not until the turn of the twentieth century that the seeds sown by Darwin came fully to fruition in two directions. Naturalists with an interest in behaviour became concerned with how animals perceived the world, and with the stimuli to which they reacted. They established the science of ethology (see Chapter 6, Signs and Signals). Meanwhile, in laboratories chiefly in the USA, a group of people revealed the nature of learning and intelligence, and founded the discipline of experimental psychology. Both were great forward leaps in the understanding of animal behaviour.

A QUESTION OF LEARNING

During this century, blue tits have learned to open milk bottles

On an island off the coast of Puerto Rico, crowded with monkeys, a mother rhesus macaque puts her infant down and walks away from it. On her part, it is a marked change of behaviour towards her offspring. Since birth, she has kept her baby close and secure. On those occasions when the little monkey has attempted to explore beyond the security of her body, she has given it freedom to wander only at arm's length, often tightly clenching its tail like a lead, sweeping it to the safety of her bosom whenever she or her infant became anxious. Now, she sits a few paces away from her terrified, squealing baby, watching its desperate attempts to follow her. When it has almost regained the haven of her body, she suddenly gets up and circles the distressed little creature, prodding and poking at it in an attempt to keep it moving. After a minute or two, she scoops up her baby, and, clutching it to her belly, rejoins the rest of her troop.

Despite its callous appearance, the 'teasing' had a serious purpose. The mother was teaching her infant to walk. The enforced exercise promotes the development of the muscular strength and control so vital to a monkey which needs to scamper confidently through the trees. We take for granted the fact that monkeys are 'born' acrobats, that cats are 'natural' mousers, and that squirrels take to cracking nuts like ducks to water. Behaviour patterns like these which are designed to assist the survival of animals in the wild world can arise through inheritance, as a result of learning, or through a combination of both processes. Unfortunately, the people who tried to discover what animals learn, and those who sought to establish what they are born with, isolated themselves into two

Soft, cuddlesome and warm mothers are the best of all. Without social contact baby monkeys develop into disturbed adults

camps in a rift which persisted for fifty years. The argument was over the relative importance of nature and nurture.

The origin of the conflict can be traced to the end of the nineteenth century, when two sets of people focused their attention on animal behaviour: zoologists on the one hand; those interested in psychology and education on the other. Darwin's belief in 'mental continuity' between man and beast had already set the stage for the appearance of experimental psychologists, who employed animals as simple models of human learning. It assumed that animals placed in test situations might illuminate the ways in which people conquer their problems. Towards the end of the nineteenth century, there was no shortage of evidence purporting to demonstrate that animals were capable of impressive feats of cleverness. A flush of vaudeville acts – dancing bears, talking dogs, counting horses – seemed to prove that of course animals could perform human-like tasks if they were instructed correctly. That was the view of Wilhelm von Osten, whose erudite stallion became famous the world over for his mathematical prowess.

A Prussian aristocrat by birth, von Osten prided himself on his abilities as a teacher. He reasoned that animals perform mentally less well than man, because they are not educated as thoroughly as human children. In 1888 he had a chance to demonstrate the power of instruction. He inherited sufficient wealth to enable him to buy a magnificent house in Berlin, in which he could devote himself wholeheartedly to his experiments. Purchasing a horse, he single-mindedly set about teaching it as though it were a human infant in the classroom. By using skittles, an abacus, and blackboard, he schooled the horse in basic arithmetic, rewarding the animal with a bite of carrot when it gave the correct answers to his questions. Unfortunately, his first pupil died, but he was more fortunate with his second one. This was a Polish stallion, christened Kluge Hans (Clever Hans). Two years of valiant perseverance were rewarded, as Clever Hans emerged as an equine genius, eagerly tapping out with his hoof the solutions to problems. What was more surprising, he invariably gave the right answers. Lesson followed lesson in relentless succession. If the horse's attention seemed to falter, von Osten could be heard by his neighbours shrieking abuse and threatening to use his whip in a manner worthy of a Dickensian school teacher!

Soon, rumours began to spread about the 'wonder horse', and many people came to watch it demonstrate talents worthy of a twelve-year-old child. Clever Hans' reputation even reached

America. Some people were sceptical, and considered the whole business nothing more than a hoax or, at the best, a highly polished vaudeville act. Von Osten, however, remained unshaken in his belief that he possessed a 'thinking' horse. Anxiously he sought scientific corroboration. After petitioning his Emperor, Wilhelm II, his request for a proper investigation was eventually granted.

On 6 September, 1904, a commission of thirteen people visited von Osten's stable yard. In charge was the director of the Berlin Psychological Institute, Professor Carl Stumpf. He had chosen the members from various backgrounds, on the grounds that not everyone in a multi-talented group could be deceived. It included a famous circus trainer who would be able to detect any signs of show-business trickery; there was also a zoologist – Dr Oskar Heinroth; a vet; and a politician. For several days, von Osten put Clever Hans through his paces. The horse was on good form, providing the answers without hesitation, and receiving well-earned titbits of carrot. The dignitaries were frankly amazed, and were unable to detect the slightest sign of fraud, or even of involuntary cueing on the part of von Osten. Perhaps his coaching, modelled on the techniques of primary school teaching, had after all registered in the horse's mind. The commission recommended a further 'earnest and thorough scientific investigation'.

Professor Stumpf took up the challenge, and promptly offered the project to one of his most intelligent students, Oskar Pfungst, who devised a series of tests designed to explore the limits of Clever Hans' understanding of arithmetic and language. It was an exemplary study, which eventually destroyed von Osten's reputation.

Pfungst confirmed that Hans was indeed 'clever', but not in the way his master had believed. Pfungst first addressed himself to discovering whether the horse could give correct answers if the questioner did not know the solution. Pfungst achieved this by conducting a test in which von Osten held up a series of numbered cards, one at a time. Normally Hans would 'read' each figure as it was presented, and correctly tap out the number on the card. In this instance, Pfungst asked von Osten to look at some of the cards before showing them to the horse, but others he was instructed to hold up without glancing at them himself. When the questioner had no knowledge of the figures he was displaying, Clever Hans' performance went badly wrong. Pfungst drew the inevitable conclusion that von Osten was unwittingly supplying the answers to

Clever Hans. The question was, how? Pfungst fitted the horse with blinkers and got von Osten to stand by its flanks, so out of the animal's restricted line of sight. From this position, the horse was asked to count. Despite the fact that it could clearly hear its interrogator, it was flummoxed. If the same commands were shouted by von Osten as he stood in front of the horse, in full view, the correct answers were tapped out. There was little doubt in Pfungst's mind that the horse was picking up visual clues, not vocal ones. He therefore switched his attention from Clever Hans' behaviour to that of his owner. Further research revealed that von Osten was 'controlling' the horse's counting behaviour by almost imperceptible alterations in his own body posture. During the training period, Clever Hans had learned to paw the ground when von Osten's head inclined slightly forward to get a better view of the hoof. In anticipation of the correct 'answer', he unconsciously tended to straighten himself, and that was the clue the horse took to stop. Pfungst found that even a slight elevation of von Osten's eye-brows, a subtle flaring of his nostrils, were sufficient to halt the counting.

At the end of his investigation, Pfungst conclusively demonstrated his theory. Standing in front of the horse, and without asking it anything, he made the stallion tap its hoof and stand to attention by slightly nodding and straightening his head! Although the final report absolved von Osten of wilful deceit, he reacted badly, and felt that Clever Hans had let him down. He sold him to another 'horse professor', Karl Krall, who remained firmly convinced of the intellectual potential of animals. Von Osten died in 1909, bitter and disillusioned.

Pfungst's classical analysis was the first scientific warning of the danger of reading conscious thought processes into an animal's behaviour. What may, on the surface, appear to be actions of great intelligence, might prove on closer inspection to be an animal providing, for food rewards, answers to 'questions' quite other than those apparently asked. For Clever Hans, there was a powerful incentive to detect what was required. If he reacted 'correctly' to von Osten's minimal indications, a bit of carrot materialised from his trainer's pocket!

Animals are acutely perceptive, and change their habits very speedily if they can, as a result, obtain some kind of satisfaction. Both in captivity and in the wild, every creature is surrounded by a bewildering variety of events, some meaningful to its existence, others irrelevant. In order to live and thrive, an animal has to recognise patterns in apparent chaos. Animals must make connections, seize upon cues, however

Performing bears in Hungary. With training, animals can develop habits which are not a part of their normal repertoire

Ivan Pavlov (1849–1936)

The striking coloration of these cinnabar caterpillars warns birds of their unpleasant taste. The birds soon become conditioned to avoid them

trivial, provided that they are relevant to something important: food, a mate or an enemy. This was what Clever Hans did, and in so doing deceived nearly everyone.

Pfungst's investigation was essentially about the way in which the horse associated events, and so came to modify its habits. Similar results were reported by the Russian physiologist, Ivan Pavlov. He stumbled across one mechanism for altering behaviour through his study of digestion in dogs. Strangely enough, Pavlov was not really interested in animals at all, but, in accordance with post-Darwinian medical tradition, regarded them simply as experimental substitutes for people. Yet his pioneering research had the utmost significance for the science of animal behaviour. It is often difficult to say what something *is*, until it is possible to say what it is *like*. Pavlov's curiosity was fired by the notion that the body was like a machine, with its own regulatory systems. Working in the St Petersburg Institute for Experimental Medicine, he first of all investigated how blood pressure was controlled. Later, he turned to the digestive system, which he compared to a factory, in which raw materials in the form of food are passed through a series of compartments. In each they are bathed in, and broken down by, enzyme-rich juices. Pavlov used dogs to analyse the process. He was a skilled surgeon, and was able to re-route the dog's salivary ducts so that they opened on the outside, just behind the angle of the jaw. There he was able to collect and measure the saliva.

Pavlov was determined to find out what actually caused the digestive juices to flow. He discovered that the process was implemented by the nervous system. When he stimulated certain areas of the dog's mouth by applying a dab of powdered meat, saliva immediately started to drip into his glass collecting tube, thereby in normal circumstances preparing the mouth automatically for receiving food. For this and many other demonstrations, Pavlov was awarded the Nobel Prize in 1904, when he was fifty-five years old.

In those days, relatively little was known about how the nervous system worked. What knowledge was current was based upon the ideas of an English surgeon, Sir Charles Sherrington. He defined actions, or behaviour, as a series of 'reflexes', of which the knee jerk is perhaps the simplest of all. Gently hammering the knee causes stretch receptors to transmit impulses along the nerve cables to the spinal cord, where they trigger off the motor nerves supplying the thigh muscles. A fraction of a second after the area just below the knee is struck,

the leg compulsively twitches in a reflex action over which there is no conscious control whatsoever. To Sherrington, it seemed that these 'behavioural adjustments' – or reflexes – were predetermined by the nervous wiring of the body, linking stimulus to response.

Pavlov saw in his drooling dogs the operation of a 'salivation reflex', and traced its neural pathway from sense organs in the tongue to the salivary glands. However, during his research, he began to be plagued by a habit which was as puzzling as it was annoying. When his Russian hounds became familiar with the experimental procedure, they started to salivate, not to the direct touch of meat powder, but to the arrival of Pavlov. The dogs had come to associate the sound of his footfall with the imminent arrival of a dish of food, and their mouths accordingly watered in anticipation. These premature secretions bedevilled his experiments. Pavlov called them 'psychic secretions or reflexes', and decided to investigate them. Working behind screens so that his animals could not see what was going on, Pavlov and his assistants altered their experimental technique. A moment or two before pushing a plate of meat through a hatch, they introduced a stimulus which was totally unrelated to eating, such as the ticking of a metronome. At first, the dog's mouth watered only when the food was delivered in front of its nose. However, after a number of trials, the tube on its cheek began to fill with saliva directly the metronome was set in motion. The dogs finally produced almost as much saliva to the ticking itself as they had initially produced for the food alone. Pavlov christened this new behaviour the 'learned or conditioned reflex'. Almost any kind of event would do to establish a conditioned reflex, and he ultimately had his dogs' mouths dripping to flashing lights, ringing bells, and pure tones of various pitches. Conditioned reflexes proved to be flexible, and could be both formed and destroyed. For example, if food suddenly failed to appear over the course of a few trials, a permanent-looking conditioned reflex could be discarded.

The conditioned reflex was the first example of a learning mechanism to be submitted to the rigour of scientific scrutiny, and went some way towards explaining variable behaviour. Pavlov grasped the fact that the essence of learning is to register the relationship between events – the ticking of the metronome and the arrival of meat. In the wild, birds learn to avoid the black and orange caterpillars of the cinnabar moth because they associate their evil taste with their distinctive colouration. However, Pavlov's conclusions were too sweeping.

He believed that he had discovered a phenomenon which provided the total overall explanation for behaviour. Pavlov thought that, in order to survive, every animal needed *two* kinds of reflexes: the sort that was inherited and fixed for life – like the knee jerk; and those that were acquired. These conditioned reflexes steered animals through their environments, by means of signs, sounds and smells, removing the creatures from danger and leading them to the things that they needed – like food. Aping nomenclature borrowed from physics, the conditioned reflex came to be regarded as the smallest indivisible unit of behaviour, an 'atom of action' from which could be built up far more complicated repertoires. All kinds of habits arising from training, education, and discipline, he set down as 'nothing but a chain of conditioned reflexes'. He could be forgiven for holding such a sweeping view of learnt behaviour because it was a reflection of his own distaste for the contemporary state of behavioural science. Naturalists and zoologists were still inclined to use as evidence of animal intelligence anecdotes of the kind peddled by George Romanes. Psychologists were deeply biased by abstract philosophical considerations concerning human consciousness, and were hopelessly bogged down by their own experimental method of 'introspection', pioneered by Wilhelm Wundt in Leipzig. The aim was to explain behaviour by referring to the thoughts, feelings, and images which flitted through the mind while under interrogation. By documenting these in the greatest detail, psychologists hoped to tease out into the open the 'constituents of consciousness' – the 'atoms of the mind'. It is little wonder that Pavlov preferred to analyse behaviour by measuring spittle! Reacting to the woolliness of introspection a number of scientists began to look at the problems more clearly. Pavlov was one, but there were others, including a young American, Edward Lee Thorndike.

During the closing years of the last century, Thorndike became intrigued by the speed at which animals could learn. For his research programme at Harvard, he used chicks in large numbers. At first, they got him into trouble. One landlady was so upset by the cheeping of the chicks and by the mess they generated, that she summarily dismissed him and his menagerie from her boarding house in Boston. William James, the eminent psychologist and philosopher, was sympathetic towards his work, and found space for it in the basement of his home, much to the delight of his children, who loved to help Thorndike conduct his strange experiments.

Thorndike had a penchant for observing what animals actually did, and for finding activities that could be measured. In the first instance, he used mazes as research tools, building them out of hefty books, of which there was no shortage in the James house. In the centre, he placed a chick, and timed how long it took to find its way out to where a bowl of seed was placed. At first, the confused creature took an eternity to navigate the alley-ways between the up-ended volumes. However, with successive trials the chick's performance improved steadily until eventually it scurried to the seed dish as fast as its legs could carry it.

Thorndike's experiments differed in one important respect from those of Pavlov. The Russian dogs were always rewarded, whether or not they salivated. Thorndike's animals had to perform the 'correct' behaviour in order to reach their food. This also proved to be a much more effective technique for changing behaviour, as his puzzle boxes demonstrated.

Thorndike designed a series of boxes which required an imprisoned animal to tug cords, press levers, and lift catches in order to escape. His victims were thirteen cats kept in a state of 'utter hunger' to make them all the keener to escape. One at a time, they were placed inside the boxes to see what they made of them. At first, the animals were dreadfully distressed by their confinement, and desperately thrashed around, trying to squeeze through the bars, clawing and biting at everything in sight. After several minutes a frantic cat usually contrived to operate the crucial latches by accident. The door opened, and it was confronted by a saucer of wholesome fish. Thorndike noted that practice appeared to make perfect; he watched one veritable Houdini learn to release itself within five seconds, after only nineteen trials. Some people were – and still are – convinced of the ability of cats to assess tasks and coolly to solve problems with deftly placed paws. Thorndike was unable to see evidence of 'flashes of insight'. He observed the solutions to his puzzle boxes emerge from feline panic. The appropriate responses first appeared purely by chance. They then became stamped into the cat's repertoire because they were followed by a change for the better in the animal's circumstances. Food and freedom 'reinforced' the successful acts, making them more likely to recur.

Thorndike refused to accept the notion that an animal like a dog or cat anticipates the consequences of its behaviour. That would endow his experimental creatures with mental processes that he could neither prove nor disprove. It was therefore

Learning to fish

Grizzly bears soon learn by trial and error the best method of catching salmon

Splashing about is often unsuccessful

Keeping still, with head underwater,

Is a better way of obtaining a meal

scientifically expedient to ignore such possibilities. He explained the increase of efficiency in the performance of his cats as due to the growing connections in the nervous system, which consolidated the neural routes between stimulus and response. Once established, the escape behaviour certainly had all the appearances of one of Sherrington's automatic 'reflex arcs'. For example, once 'trained', Thorndike could place a cat in a puzzle box with one of the sides removed, and it would still go relentlessly through its escape routine instead of simply walking out through the opening! Silly and uneconomical behaviour of this kind had all the elements of the sort of acts of personal superstition with which we are all familiar.

Thorndike's cats helped him to formulate a 'law of effect' which stated that 'behaviour changes because of its consequences'. It was not intended to be an abstract pronouncement, but to be applied to the way in which wild animals form habits. He believed that the 'law of effect' was a general law of nature, and that every animal is born into a world which sets the puzzles. For example, a fox learns by trial and error how to enter a chicken run, in the same way that Thorndike's cats had to deal with his specially-built puzzle boxes. The world is a variable and changing place. He reasoned that, over the course of a lifetime, the 'law of effect' selected habits which enable an animal to capitalise on a wide range of situations.

Opportunists, like grizzly bears, are ever testing their surroundings in their quest for food, turning over stones, moving logs, nibbling this, and tasting that. Should a particular action yield something edible, the reward will ensure that the behaviour returns, to the bear's benefit. Nowhere is this better illustrated than in Alaska when salmon are running up the McNeil River. In spots where the river narrows and cascades over a series of rocky shelves, parties of bears gather to gorge themselves on the fish. Fishing is not their primary activity, so they have to learn, and to refine, their own methods for taking migrating salmon. The cubs and young bears become very excited by the sight of leaping salmon, but have no style when they try to catch them. They splash around almost at random, rarely making a killing. Older and more experienced bears have learned to be more economical of effort. Some station themselves on rocks overlooking pools into which they plunge with great panache when fish come into view. Others waste even less energy, letting the salmon come to them. The bears sit by the side of narrow torrents up which the gravid salmon have to struggle. All the hungry grizzly has to do is to hook a passing fish out onto the bank with a flick of its paw. The occasional bear has learned to brave the river's surge, and stand with its head down, water flowing over its neck, until an unfortunate salmon happens to blunder into its jaws.

Even the most arresting acts of apparent cleverness can be accounted for by the process of learning by trial and error. In Miami, a green heron regularly goes fishing by using bread as bait. Holding a piece of bread in the tip of its bill, the bird dangles it in the water until a fish comes within stabbing distance. At first sight, it looks as though the bird has learned the technique from local anglers. However, an explanation based upon Thorndike's law of effect is equally plausible, if not more so. People regularly feed the ducks and fish which share the herons' haunts. Although the bread is not part of the herons' normal diet, the birds are often tempted to investigate floating crumbs and crusts. In the case of the green heron, the habit of dabbling with morsels of bread may have been rewarded by the appearance of fish, which were attracted to the crumbs. Dropping the piece of bread, the heron undoubtedly speared its proper prey. The agreeable consequences to the heron of seizing bits of bread increased the chance of the habit reappearing. The heron therefore learned to fish like an angler. In the same way, the first chimpanzee may have accidentally discovered the utility of a tool by placing a twig in a hole to find, when he pulled it out, that it was covered with tasty termites.

Thorndike did not publicise his results widely, nor did he broaden the scope of his animal research to embrace human behaviour. Instead, he changed tack, immersing himself in educational studies. It was left to another North American, John B. Watson, to develop the practical approach pioneered by Thorndike. He described a theoretical framework of experimental psychology, called Behaviorism. It has proved to have an important influence on the shaping of our attitudes to the problem of how animals learn. Watson believed that if behaviour was ever to become as precise a science as physics, animals had to be reduced to the status of 'black boxes', and regarded as 'stimulus-response machines'. It was possible to measure the stimuli applied to an animal, and to measure how it responded. What went on in its mind was immaterial, if ultimately the approach enabled scientists to predict and control its behaviour. Behaviorism was the very antithesis of George Romanes' brand of science, which imbued animals with quasi-human mentalities.

Overleaf. North American bison in a natural sauna. Bison learn to keep warm and to seek food in areas around hot springs

John B. Watson (1878–1958)

The Hampton Court Maze

Watson's origins were unusual for a scientific luminary. He was born in 1878 in Greenville, South Carolina. His father was a drunkard, much given to lechery. His zealous Baptist wife believed that cleanliness in all things was next to Godliness, and, true to her convictions, had her son John toilet-trained three months after birth! In 1891, Watson *père* left home to live a dissolute life with two Indian women. Watson *fils* took to the streets of Greenville and indulged in 'nigger bashing', brawling, and shooting, which led him into conflict with the police. At the age of sixteen, a criminal future clearly in store, he reformed almost overnight, and decided to get himself an education. A local university took him in, and he became an obsessive student. In the course of his studies, he apparently kept a 'house of rats', and taught the animals various tricks. After graduating in 1899, he raised his academic sights, and with the help of a scholarship, reached Chicago in 1900, where he met Professor Jacques Loeb who also thought of animals as 'mindless machines'. A year later, Watson had to decide what subject to pursue for his thesis, and chose 'animal education'.

Watson picked rats for his research, and attempted to discover what, and how, they learned. Although hardly the animals to endear themselves to a wide public, Watson was fond of these rodents, spending long hours observing their behaviour. Realising that they were underground animals, perfectly at home crawling around burrows and sewers, he decided to test their performances in mazes. By then, the maze was a standard research tool for probing the mental processes of animals. It had been established by another American, Willard Small, who built increasingly elaborate ones, culminating in a magnificent model based upon the famous Hampton Court Maze. Small's mazes were designed to provide problems for famished rats, and as they wrestled with the difficulties posed by the junctions and blind alleys, he speculated on their states of mind.

Something as complex as the Hampton Court Maze was far too large for Watson's purpose; it had too many choice points, and the behaviour of the rats defied analysis. Watson therefore made very simple ones – 'T' mazes with one choice point – and used them to see how habits developed. At first, his rats dithered and learned haphazardly. If they were not especially hungry, they even played and went to sleep in the passages. Eventually they caught on, and performed automatically 'like a reflex'.

Early on in his career, Watson found he possessed flair for

Fixed habits

A simple test
shows that sooty
terns home in on
their nest site, not
on their egg

A new nest is
created nearby

The egg is removed
and placed in it

But when the bird
returns it continues
to sit on its now
empty nest in full
view of its own egg

studying wild animals, and soon he was using techniques that foreshadowed those developed by scientific field naturalists in Europe thirty years later. During the summer of 1907, he spent three months on Bird Key, one of the Tortuga Islands off Florida. It was a cramped coral island barely a quarter of a mile across, and frequented by countless thousands of nesting noddies and sooty terns. Although Watson described the displays of the sooty terns in great detail, his real interest lay in discovering how each individual managed to find its own nest among such a seething mass of birds. Watson wondered whether the terns were able to recognise, and so home onto, their own eggs. To test this, he selected a nest, made a fresh scrape in the sand thirty centimetres or so away and transferred the eggs into it. To his astonishment, when the tern returned, it went without hesitation to the spot where its original, but now empty, nest was located. Further trials confirmed the fact that the birds homed onto geographical sites, using landmarks such as tufts of grass and bushes to fix the position of their nests. As a part of the study, he noticed that many birds always landed some distance from their nest scrapes, and then took a tortuous, winding path around the tangle of fallen branches which littered the ground. He cleared some of the obstructions away, but the terns continued to alight at the same places, and waddled home by the same twisting route as though weaving their way through a maze of branches. The seabirds had developed fixed habits of the kind that Watson had observed in his rats. What he had seen on the Tortugas confirmed that behaviour observed in the laboratory was not totally artificial. In any case, Chicago was more comfortable, and rats were far more controllable than free-flying sooty terns!

Watson's rats showed him that, once acquired, habits were surprisingly resistant to change. He trained white rats to run the full length of a three-metre-long alley – a straight 'maze' – in order to obtain food at the far end. The alley was then blocked half way along, and the food placed just in front of the partition. When released into the passage, the rats sped along it as they usually did and battered their noses repeatedly against the wall as though they were blind. Conversely, rats which were conditioned to run half way, with the partition in place, refused to go further when it was removed. Instead, they circled round and round in a bewildered state at the half way stage rather than scamper a little further for their reward. Like the sooty terns, the rats rigidly adhered to a habit, ignoring changes in their surroundings, at least in the short term.

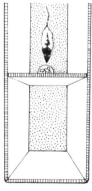

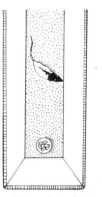

A rat conditioned to obtain food half-way up a 'maze' (above) continues to go to the half-way stage even when the barrier is removed and the food placed at the far end (below)

In 1908, Watson was appointed to the chair of Psychology at Johns Hopkins University, in Baltimore, where he introduced his ideas about Behaviorism. He envisaged a science that would prepare men and women to understand, predict, and control their own behaviour. Re-reading Darwin had crystallised his views that men and 'brutes' were fundamentally similar, and could therefore be studied by using similar methods. Lambasting the adherents of 'introspection', he pointed out that they had made no real progress in understanding human behaviour, whereas his own experiments had elucidated many aspects of the learning process. His outspokenness made him many enemies. Eventually, in 1913, he delivered his manifesto for Behaviorism, proclaiming that those interested in human behaviour should abandon their meditative techniques, and search instead for general laws of learning. These could be applied to men and women.

Practising what he preached, Watson devoted himself increasingly to studying people. For example, he analysed in behavioural terms the relationship between mothers and their babies, and was partly responsible for creating the convention in which parents treated their infants as though they were miniature adults. However, his academic aspirations were prematurely destroyed owing to an affair he had with one of his students. Watson's wife was a severe woman, not unlike his own mother, and she had procured by devious means some of his love letters. Her brother photographed them, intending to blackmail Watson into giving up his young mistress. When this ploy failed, some of the letters were circulated among his colleagues, causing such a scandal that he was asked to resign. In 1921 he started a new life as an advertising executive on Madison Avenue with the J. Walter Thompson organisation. Applying what he had learned from animals, he transformed the face of advertising, devising the notion of the 'brand image', which is still the essence of sales promotion. Taking over the Johnson's Baby Powder account, he sold it as a symbol of purity and cleanliness which every caring mother should wish to possess, and should feel guilty without! To him, people were like 'black boxes' — it did not matter what they thought, providing they bought the goods. As his rats had become conditioned to run through mazes for food rewards, so he conditioned consumers through clever advertising to accept all sorts of prejudices, and to purchase products which gave them a sense of satisfaction as they handed over their dollars. It was a measure of Watson's success and his belief in the universal

Chafing
For a young baby doctors require this special care

THERE is no fragrance in the world more appealing than that indescribable sweetness — part just clean babyness, part soft little woolens, part delicate powder—which makes a baby's skin so adorable.

But that tender skin is a responsibility. It needs the most faithful care to save it from the misery of chafing and other eruptions which assail it.

After the baby has been bathed and dried with a soft towel, powder should be sprinkled in all his little folds and creases to absorb the last stray bit of moisture. And every time he is changed he must be liberally powdered.

There is nothing—except the nipple of his bottle—which comes into such intimate contact with the baby as the powder. One cannot be too careful in its choice.

The powder should be exquisitely fine, soft, absorbent and pure beyond question.

Mothers Trust this Powder

Because Johnson's baby powder has these qualities in such superlative degree, more mothers choose this than any other powder.

One of Watson's advertisements for Johnson's Baby Powder

principles of learning, that he eventually became a very rich man. He was also the only person to have studied animal behaviour who has made a lasting impact on our daily lives.

Watson was the Messiah of Behaviorism. He formulated the creed that nearly all behaviour is the result of conditioning, and that our environment shapes our behaviour by reinforcing habits. It was left to Burrhus F. Skinner to become its chief apostle.

Unlike Watson, Skinner had no particular desire to watch animals. He was simply curious to discover as much as possible about how behaviour changes. To him, animals were a means to an end. While at Harvard, he perfected a device which saved him the labour of observing animals and measuring their behaviour. It was rather like an automated puzzle-box, in which a confined animal could record its own activity. By learning to press a lever, a rat or pigeon obtained a small pellet of food or switched off a mild electric shock delivered to its feet. Once the apparatus was set up, no one needed to watch the creature because its lever-pressing activity was displayed on ticker tape. For a scientist the Skinner Box, as it was called, had many advantages. Serried ranks of them could provide huge amounts of information which could be collected by an untrained assistant.

Skinner made some surprising discoveries through the use of his boxes. For example, he found one day that he was short of food pellets, and needed to ration the rewards. He programmed one of his boxes to dispense a pellet after a bout of lever pressing rather than after each press. Skinner predicted that the pigeon would lose interest because the food would appear too infrequently. He was wrong. The bird worked even harder for its spaced-out rewards, attacking the bar with enthusiasm. This experiment led him to realise that the scheduling of food rewards had a profound effect upon the speed of learning. Lever pressing was most heavily reinforced by infrequent rewards. Gambling is the human counterpart. The prospect of an occasional win on a fruit machine is sufficient to cause addicts to pump the handle as though their very lives depend upon it.

If a delay of only a few seconds was introduced between an animal pressing the lever and the appearance of a pellet, then the learning process was greatly slowed down. Skinner discovered that, in these circumstances, performance could be restored if a further stimulus was introduced to bridge the gap between the lever pressing and the delivery of a pellet. Skinner

Behaviour inside a Skinner Box. The rat learns to press a lever to obtain a reward when the light shines

used a flashing light, or a sound. The technique was called secondary reinforcement, and was already in use among circus trainers in order to produce virtuoso performances from their animals in a situation where rewards cannot be administered promptly. Under the Big Top, the blast of a whistle enables the animal stars to make the connection in their minds between doing what is required of them, and the signalling of an imminent presentation of sugar or carrot. In every-day life, secondary reinforcement is a widespread phenomenon. Money is a powerful secondary reinforcer; although useless in itself, it allows us to reap the rewards of our labour – food, housing, clothing, and so on.

Behaviorism provided potent techniques for manipulating behaviour and promoting learning. They were – and still are – successfully used to cure aversions, phobias, and fetishes. By applying a regime of rewards for desirable behaviour, animals could be made to act with amazing inventiveness. Skinner's students taught pigs to play pianos, parrots to ride bicycles, and chickens to dance at the drop of a grain of corn. In Boston, Massachusetts, Mary Willard has put capuchin monkeys to work as home-helps for seriously disabled people. Using Skinnerian techniques of behavioural shaping, she has trained her animals to follow a spot of red laser light around their owners' apartments, switching lights, collecting drink from the fridge, dusting the carpets with specially designed vacuum cleaners, and changing gramophone records. Every monkey is taught to spoon-feed its paralysed owner, comb his hair, and to turn the pages of a book when he says so. The animals obligingly carry out their tasks for the reward of a peanut dispensed from a container fitted to the invalid's chair.

During the decade before the Second World War, Skinner felt that he could account for virtually the whole spectrum of human and animal behaviour. He argued that we are all born equally into the world with 'minds like blank slates', which then become scored with habits fixed by the 'contingencies of reinforcement'. In other words, we all become creatures of our conditioned responses. Even something as fundamental as the attachment, or 'love', for its mother shown by a child was seen as behaviour shaped initially by the provision of frequent rewards of breast milk; during feeding, the baby associated the mother's smell, looks, and voice with the satisfaction of swallowing the warm, nourishing liquid. The environment was the all-important provider of rewards and punishments, and this was ultimately what shaped behaviour.

Home help. By applying a regime of rewards, a capuchin monkey can be trained to assist disabled people

Skinner was not without his critics. Even from within experimental psychology itself, there were those whose opinions were at variance with the behaviourists. Early in this century, L. T. Hobhouse, a sociologist and journalist, devised a series of tool-using tests for some of the monkeys and apes in the Belle Vue Zoo, Manchester. These called for the animals to reach food by placing boxes in strategic positions, or by manipulating sticks. He was very impressed by the performance of some of the individuals, especially of a chimpanzee called 'the professor', which solved problems so rapidly that it appeared to have flashes of insight. In 1917, Wolfgang Köhler, a German working on the island of Tenerife, confirmed Hobhouse's results. He watched chimpanzees using poles to rake in bananas placed beyond their reach, fitting sticks together to make long tools, and constructing ladders out of piles of boxes. This and other evidence led Köhler to suggest that a solution to a problem may possibly appear in just one step – as a 'picture in the mind or *Gestalt*' – rather than in the series of stimulus-response steps envisaged by Skinner.

Skinner's sternest critics were zoologists. Their traditional interest lay in the distinctions between species. They believed that the behaviour they saw in the wild had the utmost survival value, and was therefore shaped by evolution, not by regimes of reward. This meant that habits had a genetic basis. By contrast, the Behaviorists stripped away all the characteristic differences between animals. To them, a rat was reduced to a lever-pressing robot, and a pigeon likewise became a button-pecking machine. They were performing essentially identical tasks, and so there was little to choose between them in the laboratory. Had Behaviorists watched how their animals passed the time in their holding cages and Skinner Boxes, they might have been less dogmatic in their assertion that all behaviour is learned.

Every rat cleans itself, indulges in mutual grooming, hoards food, squabbles, courts, and mates. Every pregnant female rat constructs a nest if she can lay her paws on suitable material. In it she gives birth to her pups. She is a good mother, and gives the impression that she knows exactly what to do. Laboratory rats vigorously lick their pink and naked babies beneath their tails to make them pass urine. By the time they reach the age of three weeks, the pups are frisky, playing boisterous games of tag. Even within the limits of the Skinner Box experiments, a rat's behaviour is much richer than the automated record admits. Some press the lever with their forepaws, others butt it

One of Wolfgang Köhler's chimpanzees using boxes and a stick to reach food

The skill of the weaver bird is an example of behaviour that stems from instinct rather than learning

with their noses. Between the bouts of lever pressing, they often groom themselves with a vigour that exceeds normal toilet behaviour, and go through the motions of building a nest, although no straw is present. Such acts are born of frustration. Zoologists like Konrad Lorenz, who observed geese on the banks of the Danube, had for a long time doubted whether behaviour patterns like those were generated by the process of Skinnerian conditioning. They were inclined to believe that nature was at the root of more behaviour than the Behaviorists cared to concede.

The scientific study of how birds orchestrate their voices undermined the entrenched positions of those who saw *either* instinct *or* learning as the basis for behaviour. Some species, like cuckoos, have songs which are monotonous in their regularity and appear not to be learned. They are produced with the mechanical precision of an inborn 'reflex'. Some elaborate bird music may also be determined by inheritance. The liquid warbling of roller canaries may be a case in point, because these birds have been bred by generations of fanciers for the beauty of their voices. But there is circumstantial evidence that some birds learn their songs, at least in part. Every starling and mockingbird sings differently, and parrots and mynahs are veritable mimics.

The subtle role of experience and inheritance in the development of bird song was demonstrated in the heart of the English countryside, where William Thorpe became interested in the voice of the chaffinch. The chaffinch, a common British woodland species, has a relatively simple song lasting about two and a half seconds. Yet it sings with recognisable regional dialects throughout its range; for example, a bird resident in London sounds different from a Scottish chaffinch. Thorpe believed it unlikely that the diverse accents owed anything to genes; more likely they were caused by local traditions. Working in the post-war years at the Madingley Ornithological Field Station just outside Cambridge, Thorpe's objective was to discover just how much of the chaffinch's song was learnt, and how much was ingrained in its genes.

It was onerous work. Each spring and summer, Thorpe and his assistants scoured the woods and hedgerows around Madingley for the nests of chaffinches, and, in the interests of science, stole the nestlings when they were five days old. In the laboratory, they proved to be demanding guests, and had to be fed twice hourly throughout the day on a nourishing potpourri of biscuit meal, ants' eggs, seeds and cod liver oil.

Young cock chaffinches may learn some of the details of their father's song while still in the nest

Sound spectrographs of a wild chaffinch's song (above), and of an isolated bird (below). After Thorpe

One of the toughest challenges faced by anyone studying bird voices was how to 'see' and measure the trills, flourishes, and modulations which flow in rapid succession from a bird's throat. Daines Barrington had discovered two centuries before that musical notation failed miserably to capture the details. By 1950, technology had come to Thorpe's help. The introduction of the portable tape recorder obviated the need to drag cumbersome disc-cutting machines into the woods. Most importantly, the Bell Telephone Company had invented an instrument called a sound spectrograph. This revolutionised the business of analysing bird voices, doing for sound what the prism does for light. It digested any noise, sorted out all of the frequencies moment by moment, and regurgitated the information as a pattern of peaks and troughs. Thorpe saw one of these machines in the USA, and discovered that there was only one in the United Kingdom. It was held under lock and key by the Admiralty, who were rather upset to receive Thorpe's request to borrow the instrument because they regarded it as top secret! Nevertheless, they finally relented. Without the sound spectrograph Thorpe's pioneering research would have been impossible, because he needed to tell at a glance the difference between the songs of birds with different experimental backgrounds.

To tease out the inborn elements of the chaffinch's song, Thorpe adopted the time-honoured technique of rearing individuals in solitary confinement. He took nestlings and installed each in its own private lead-lined, sound-proofed box so that it was unable to hear other birds sing. During the following spring, each of the orphans opened its beak and uttered crude, harsh rattles, which were approximately the correct length and consisted of about the right number of notes for a chaffinch's song. There the similarity ended. These primitive songs were abnormally simple, and lacked the tonal purity and the four distinctive phrases of the fully-developed song of the species. This was plain from the pictures produced by Thorpe's wonderful spectrograph. This ungainly warble he interpreted as the genetic blueprint – the song which comes 'instinctively' to a cock chaffinch which has been deprived of the opportunity to hear, and learn from, its fellow choristers. The questions which now needed to be answered were when and how do young chaffinches learn to refine this inborn rattle into a proper song?

Thorpe played songs of various kinds of birds to another batch of orphaned chaffinches before they were able to sing themselves. On the whole, they ignored them, confirming the conventional wisdom of aviculturists that the chaffinch is a poor mimic. Nevertheless, his birds proved to be very receptive to the recorded songs of their own kind, showing that their ears were innately tuned towards chaffinch language. This built-in bias determines the sort of sounds they will emulate.

Thorpe also assembled groups of hand-reared birds so that they could listen to each other, but prevented them from hearing the voices of wild adult cocks. To his surprise, they eventually composed fairly tuneful songs; not perfect, but certainly better than those of completely isolated individuals.

Thorpe's most astonishing discovery was that completely isolated immature chaffinches only had to hear an adult sing for a brief period of a few days during the autumn in order to develop nearly normal song patterns the following spring. In the countryside, young chaffinches probably hear their fathers vocalising and have plenty of time to absorb the details they need to polish their own singing when it is their turn to stake out their nesting territories. Thorpe's captive birds passed through a 'sensitive period' during the autumn when they were capable of committing to memory the details of true song. These they then used to add musical flourishes to the rough vocal prototype with which they were born.

While Thorpe was proving that early experience could influence the outcome of adult behaviour, there was growing evidence also from within the psychologists' camp that behaviour may be the result of instinct as well as learning. During the 1940s, Skinner was at the height of his influence, resolutely maintaining that animals were pushed this way and that by the 'contingencies of conditioning'. In shaping behaviour, the influence of food and other rewards played the dominant role, he claimed. Lorenz, with his 'imprinted' geese, had already shown that the bond between a gander and her goslings was forged independently of food rewards. This observation had little effect upon psychological thinking. But the discoveries of Harry Harlow, a Skinnerian-trained psychologist working at the University of Wisconsin, could not be ignored.

Harlow was investigating how young rhesus monkeys learned. He instituted a controlled breeding programme to supply him and his colleagues with a steady supply of healthy infants, each of which had the same rearing history. Mother monkeys were rather a nuisance because they were idiosyncratic, so he dispensed with them as soon as possible. The infants were separated from their mothers shortly after birth, and raised in their own individual cages. Each was given a nappy – a cheesecloth diaper – which served as a security blanket. The little monkeys became very attached to them, and were frantic when they were removed for sanitary reasons. To his surprise, all Harlow managed to produce for several years was dirty towels and distressed monkeys! When the females grew up they became atrocious breeders. Harlow had a little trouble in the first place in getting them pregnant. Without having experienced the warmth of a real mother, the severely deprived females proved to be reluctant sexual partners. Some were eventually forced into motherhood through the patience and enterprise of Harlow's stud monkeys. When finally they gave birth, the motherless mothers were unable to conjure up a grain of affection for their babies. They treated their squealing, clinging infants as though they were parasites, wiping them from their bodies and occasionally thrusting their tiny heads into the wire floors of the cages. The conclusion was inescapable: early balanced company, with all of the experience and stimulation it brings, is necessary for the expression of normal primate parenthood.

To solve the problem, Harlow in 1957 gave his simian orphans surrogate mothers: wire frames covered with cloth and equipped with baby feeding bottles. These were an enormous success. Harlow decided to analyse the nature of the mother-infant relationship (he used the word 'affection' with a healthy eye to the publicity spin-off!). Using various kinds of model mothers, Harlow conducted a series of experiments to explore the *instinctive* preferences of the baby rhesus monkeys. During the first fifteen days of life, they rejected cool cloth models in favour of warm wire ones. Up until the age of six months, they chose gently rocking mothers rather than static ones. More recently, lactating cloth surrogates have been constructed at Wisconsin, with ice water pumped through their 'veins'. The little orphans abandoned them and remained aloof from their model mothers forever. With characteristic humour, Harlow observed wryly that there was only one social affliction worse than a frigid wife, and that was an ice-cold mother!

His experiments also undermined a fundamental belief of psychology. When given a cloth mother without a milk dispenser, and a hard wire lactating model, baby monkeys invariably went to the soft one to cling to when startled. Although it could provide no nourishment, the soft, milkless one became the centre of attraction. Traditional psychologists would have predicted that the lactating model would always have offered the best reassurance, on the grounds that all a baby desired

Learning to walk

A mother rhesus monkey teaches her baby to walk. She calls to it

And prods it

With more encouragement, it gets to its feet

And walks over to her

was a suckling mother, no matter what her tactile properties. Harlow's demonstration that baby monkeys wanted to cuddle as well as drink milk caused experimental psychologists to re-examine their theories. As he stated, 'There is more than merely milk to human kindness.' Early 'affection' was clearly important. A baby monkey isolated in a strange room with only a bare wire surrogate mother for comfort, became a pathetic little creature, cowering in the corner, clasping itself and screeching with terror.

But even those babies which were given warm, cuddlesome surrogates grew up to be socially awkward, relative to monkeys with natural upbringings. The passive model mothers were unable to initiate play, or to encourage exploration. The orphans reared on surrogates were, however, better off than those that were *totally* isolated for the first six months of their lives in Harlow's 'pitiless pit'. Each one languished alone in this stainless steel prison without even the scant comfort of a wire surrogate. When introduced to groups of normally-reared monkeys, these individuals froze in fear, or were overcome with fits of uncontrollable rage; some undertook suicidal missions by attacking dominant males, something that no normally brought up monkey would ever do. Their severely deprived upbringing had resulted in irreversible 'behavioural deficits'. Quite literally, they were mad!

Harlow made a further discovery. Motherless infants which were allowed to meet every now and then in play-groups grew into better-adjusted adults than those individuals which were closeted alone with their real parents. Clearly, parents are not as vital as everyone had previously thought in preparing their offspring for their life ahead; peers will do just as well! Needless to say, babies which were both mothered and had the opportunity to play with other youngsters, as would happen in the wild, turned out best of all. The situation was reminiscent of Thorpe's chaffinches.

It is important to realise that Harlow and his helpers were not trying to be cruel to monkeys. Working with primates was designed to aid them in thinking about human problems. Then, as now, their research programme was designed to elucidate sociological phenomena by using monkey models. To some extent, they were successful in gaining an insight into the causes of delinquency, baby battering, and depression. Harlow's successor, Steve Suomi, has devised a useful technique for curing depression. He restores socially-deprived adult monkeys by introducing them to infant play groups. It works, because severely withdrawn animals are intimidated by the presence of confident and socially robust grown ups. They react much better to babies. Young animals are socially reassuring, and depressed adults can be brought out of themselves and can learn with the infants how to strike up relationships. The method is now used in Wisconsin to nurse depressed people, using the assistance of normal, healthy children.

The interplay between nature and nurture which concerned both Thorpe and Harlow can be observed in the free-living rhesus monkeys of Cayo Santiago. The islet is essentially an open-air laboratory, continuously monitored by scientists who have, since 1938, daily visited the place to replenish the food hoppers with nutritious monkey food. There are no predators, and, apart from the occasional summer hurricane, life among the palms and mangroves has a halcyon quality. On Cayo, it is possible to witness the role of real mothers in educating their infants.

During the spring, the new-born babies clinging instinctively to their mother's fur, their tiny mouths latched onto the nipples. From this safe position, the baby is able to watch other members of the troop, and to get an excellent view of what its mother eats, where she goes, whom she grooms, and who grooms her. As each baby becomes strong and confident, it gradually makes sorties from its mother's warm and comfortable body. As we saw at the beginning of the chapter, she even encourages it to walk, something that the baby often finds rather traumatic. After a week or two, the baby ventures further afield, but regularly returns to its mother for reassurance. If the baby has an elder sister, she will occasionally take turns in fondling and grooming the infant. During this period, the baby meets others of its own age, and begins to play rough-and-tumble games of tag and wrestling. One moment an infant is bounding after his companions, next, the tables are turned. Should the boisterous juveniles irritate one of the adult males, they may incur his wrath, and be cuffed or bitten for their insubordination. That is how they learn the discipline of a monkey troop.

Their rich repertoire of facial expressions is doubtless inherited. But they have also developed habits which are copied from one generation to another — culture. On Cayo Santiago, water games are very much in vogue. The youngsters are introduced to the custom by the parents, who regularly enter the sea to splash and bathe, especially in hot humid weather. The infants, like human children, seem to love to dunk themselves, either in the sea or in fresh-water pools, and, where possible, queue up to fling themselves from favourite branches, belly-flopping into the water with tremendous splashes. The sheer exuberance of their behaviour is difficult to describe in dry academic language. It seems thoroughly enjoyable!

But babies have no future, except as adults. Perhaps the playing is a rehearsal for behaviour which the monkeys will need when they grow up. Some of the juvenile high spirits are expressed in acts which look surprisingly similar to mating. For a monkey, that always has survival value.

Over the course of the past two decades, the entrenched views of experimental psychologists and zoologists have mellowed under the influence of more moderate and pragmatic scientists, like Robert Hinde, of Madingley, Cambridge. Hinde, and those like him, have sought for common ground, and have persuaded the advocates of both disciplines that behaviour is the result of a blend between nature and nurture; most habits are learned within constraints imposed by an animal's genes. The details vary from species to species. For instance, the chaffinch never learns to use its foot for holding food; presumably it is precluded from doing so by its inheritance. This habit develops naturally in great tits, and they become increasingly adept with practice. Experience also indicates to finches what size of seed they can most economically husk, given the inherited shapes of their beaks; hawfinches can tackle cherry stones and split them with little trouble, but goldfinches rapidly learn to avoid heavy-weight nuts in favour of thistle tweaking.

This raises the question of why some animals should have a greater capacity to learn than others? This can be answered in a number of ways. In essence, some creatures live in changeable environments, and possess life styles which require them to solve problems with speed. Behaviour patterns which are firmly anchored in genetics change too slowly to cope with such daily adjustments in habits. Although natural selection has endowed creatures with inborn 'strategies' for obtaining food, procuring mates, and for avoiding predators, those species which are faced with problems which need to be solved quickly have open areas in their genetic instructions which are filled in by learning. A fairly humble, and relatively unintelligent mammal, like a rabbit, can survive on its instinctive behaviour patterns. But hunters, such as dogs and cats, whose prey can present themselves, and attempt to escape, in innumerable different ways, must 'learn' from experience in order to operate well. On any scale of 'intelligence', such mammals perform well, as do higher primates and dolphins, which, like social hunters, need to acquire the skill of collaboration.

A blue bird of paradise displaying

SIGNS AND SIGNALS

Kittiwakes disputing a nest site.
After Tinbergen

While experimental psychology was developing, great strides were being made by scientifically-inclined naturalists who took the laboratory out into the field, and the field into the laboratory. In doing so, they advanced considerably beyond Darwin's pioneering analysis of animal communication.

In the rain forest of New Guinea, a blue bird of paradise appears to take leave of its senses. Uttering a rhythmic, mechanical gurgling sound, with plumes all of a quiver, it suddenly pivots forward to hang upside down beneath its moss-covered perch. From this curious inverted position, it displays a startling black and red velvet breast patch amid a cascade of shimmering cobalt blue feathers.

Somewhere on the Russian steppes, a turkey-sized great bustard seems to accomplish the impossible by turning its plumage inside out, taking on the shape of a walking feather duster.

High up in a Canadian spruce tree, a pied woodpecker takes careful aim at a dead and resonant branch, and hammers a resounding tattoo with such fury that one can only marvel at how the bird avoids shattering its own skull.

These acts of apparent folly are part of the 'language' of birds, which in turn is an expression of the ability animals possess to communicate with each other. Passing and receiving messages is vital even for comparatively solitary creatures like, for instance, hedgehogs. Most animals must continually make alliances, maintain friendships, and sever relationships which have outgrown their usefulness. Animals also have to influence each other's behaviour; dominant individuals must maintain

Karl von Frisch, who discovered
the language of honey bees

their authority; young ones must stimulate their parents to feed them; bees must direct the flights of workers from the hive to nutritious blooms. Transactions of these and other kinds are made possible by the exchange of signals of one sort or another. The courtship of the blue bird of paradise catches the eye of a hen, whereas the woodpecker's drumming message of territorial acclamation is designed to impress his rival's ear. A bull black rhinoceros uses the breeze to deliver a similar message, drenching patches of his savannah home with scented urine. No matter the form of the 'language', signals are a barometer of an animal's mood, and an important social lubricant. They enable those who can interpret the signs to anticipate what the signaller will do next, whether it is likely to flee or fight, to behave sexually or not, to be open to overtures of friendship or likely to rebuff them.

Animals have evolved such a bewildering variety of displays that, for a long time, they baffled anyone who attempted to make sense of them. After Charles Darwin's promising start, further advances did not take place until well into the present century, when scientists began to decipher the 'languages', and to explain why displays should differ between species. It was perhaps the greatest challenge of all in the study of animal behaviour, and was firmly taken up by a bevy of gifted naturalists, mostly from central Europe. Three of them, Konrad Lorenz, Niko Tinbergen, and Karl von Frisch, eventually shared a Nobel Prize for their achievements in 1973.

Von Frisch was first in the field, and became famous for unravelling the language of bees. But his discoveries stemmed from his interest in sense organs – the 'windows' through which animals experience the world. This led him to develop a marvellous technique for framing questions about perception, to which animals themselves could give clear, unambiguous answers.

He was born in 1886 into a wealthy, cultured Viennese family. His parents had already established a pastoral summer retreat in the hamlet of Brunnwinkl, overlooking Lake Wolfgang. This was to play a vital role in priming the boy's curiosity about natural history. An old mill which the family had acquired needed renovation, and over the years, his father, Professor Anton Ritter von Frisch, sank the savings from his successful medical practice into the property. The remainder of the houses in Brunnwinkl were bought by Karl's maternal uncles, so that the place became a closely-knit community of friends and relatives during the summer months. It still is.

Then as now, that part of Austria, just east of Salzburg, nestling in the lee of the Schaffberg Mountain, was not noted for balmy summer weather; it rained far too often for comfort. However, there were compensations. The hills wore a garment of flowers, the woods rang to the songs of birds, a myriad of insects hummed in the meadows, and the margins of the lake teemed with insects, frogs, and fish. This was the world which inspired the imagination of the child from Vienna. As he grew up, Karl von Frisch's preoccupation with wildlife intensified. Whenever possible, he watched animals, noted their customs, and made pets of them. Like many budding naturalists, he developed a passion for collecting. In 1902, one of the upstairs rooms in the mill was commandeered as a museum in which to house his burgeoning store of moths, butterflies, and beetles caught in the neighbourhood of Brunnwinkl. As time went by, pressed flowers, mounted birds, and fossils were added to make his collection representative of the district.

Von Frisch entered Vienna University in 1905 to train as a doctor. But his heart was not in medicine. He was inspired most by the lectures given by his uncle, Sigmund Exner. Exner was an expert on the multi-faceted eyes of insects and crustaceans, and through him, von Frisch became fascinated by vision, and the wider topic of how animals perceive the world. After graduating, he built his scientific reputation on a series of physiological discoveries. But as time went on, he gradually became more interested in interpreting behaviour as a means of obtaining information from animals about what they could see and hear.

During the early part of the present century, the question of whether fish could perceive sound was hotly debated in zoological circles. From the design of their ears, it seemed reasonable to suppose that they were stone deaf because certain structures thought to be vital for hearing were missing from their inner ears. The leading exponent of this view was Professor Otto Korner, the Director of the Ear Clinic at Rostock University. He was in no doubt that fish were deaf and devised a number of eccentric experiments to prove his contention. He assembled tanks of fish, pursed his lips, and whistled merrily to them in as many ways as he could muster. As the fish were unmoved, he treated them to the ultimate aural experience, exposing them to *lieder*. A famous lady opera singer was hired, and solemnly regaled them with ear-shattering flights of coloratura, but the fish were no more responsive to her florid soprano trills than to Professor Korner's tuneless whistle. This

Bees can see blue. Karl von
Frisch discovered this by
a simple test (above) with
coloured squares set among
cards of various shades of grey

'concert' appeared to confirm his belief that fish could not hear!

Karl von Frisch was not convinced of the veracity of the professor's results. He knew that an animal's behaviour usually assisted its survival in some way. Putting himself in the fish's position, he reasoned that even the most heavenly singing would mean nothing unless it could somehow be made crucial to the creature's existence. A fish might be tempted to stir to the sound of music if it heralded the appearance of food? He set to work with a blinded catfish. (It was essential for the success of the experiment that the fish could not see and so react to von Frisch's presence.) Several times a day, he approached the tank, and whistled a few notes immediately before placing a juicy morsel beneath the catfish's snout. At first, it only reacted to the food, using its sensitive barbels or 'whiskers'. But after six days of this routine, the catfish connected the noise with the arrival of a food pellet – rather like Pavlov's dogs. It lunged out of its lair as soon as von Frisch whistled, but *before* he had time to offer the fish a piece of worm. The catfish therefore provided him with an affirmative answer to the question of its hearing ability. This convincing demonstration was laid on for Professor Korner, who accepted his own mistake with alacrity.

Just before the outbreak of the Great War, von Frisch was employed at Munich University, and became embroiled in a heated controversy over whether lower forms of life such as insects could see in colour; or only in shades of grey. Characteristically, he sought inspiration in the countryside around Brunnwinkl – he always felt that scientific research thrived best in beautiful surroundings, and never lost a chance to work in the relaxed atmosphere of his family's country home. In the summer of 1912, he watched honey bees from the family's hives visiting the conspicuously coloured flowers around the garden. To him, the blooms resembled inn signs vying for the custom of pollinating insects in return for gifts of nectar. It seemed inconceivable that the 'glory of flowers' should amount to nothing more than a monotonous display of shades of grey in the eyes of colour-blind insects. He had a hunch that insects had perfectly good colour vision, and turned to his bees to provide him with an answer. They gave it to him surprisingly easily.

Von Frisch trained his honey bees to visit little containers of sugar solution which were located on squares of blue cardboard. After a very short time, his bees associated the 'colour'

A bee's-eye view of a flower.
Red or purplish flowers are
strong reflectors of ultra-violet
light, which bees can see. Such
ultra-violet beacons give
promise of nectar or pollen

with the sweetness. He then took away the food and placed the blue square among a collection of other squares painted in all shades of grey from black to white. If the bees were colour-blind, they would mistake one of the matching shades of grey for the blue one, and assemble on both in search of their sugar water. When von Frisch performed the crucial test, the bees completely ignored the grey squares, flying immediately to the solitary blue square on which they had come to expect a supply of food. There was no doubt that they could perceive blue as a distinct colour.

Von Frisch used the same technique to test his bees' sensitivity to other colours, and made some surprising discoveries. Although they had no difficulty separating yellow and green, the insects always confused red with black or very dark grey, indicating that they were completely insensitive to red. However, they compensated for this deficiency by perceiving ultra-violet, a colour to which man is blind. This discovery was made in 1937 by one of von Frisch's students, Mathilde Hertz. She noticed that bees could tell a difference between squares of white paper which, to her eyes, appeared identical. When she analysed the light from them, she found that some sheets strongly reflected ultra-violet. A little further research revealed that ultra-violet was the brightest and most luminous colour perceived by honey bees.

These discoveries enabled von Frisch to appreciate how well many flowers acted as signals, beckoning bees. Botanists had already noted the rarity of purely red flowers. This was no puzzle to the Austrian naturalist; such blooms would be perceived as black and unimpressive to the insect eye. Several so-called 'red' flowers reflected a great deal of blue light and took on a mauve or purple hue to passing bees. Even the brilliant scarlet poppy was more than it seemed to the human eye. To the honey bee, it stood out as a dazzling ultra-violet beacon with a promise of protein-rich pollen – poppies have no nectaries. The majority of white flowers either absorbed or reflected ultra-violet from sun light, and therefore appeared brightly tinted to flower-hunting bees. Many flowers exhibited eye-catching patterns or spots which led the bees to the nectaries, and so brought them into contact with the sticky stigma and pollen-covered anthers: those of the forget-me-not, for instance, sat in the centre of the blue floral disc, like yellow bull's-eyes. These marks were called bee guides, and were sometimes picked out in ultra-violet – visible only to bees and some other insects.

The coloration and patterns of flowers perfectly illustrated the way in which signs and signals evolved hand in glove with the perceptual abilities of the animals to which the messages were directed. In this case, flowers which depended upon bees, were designed for their eyes only!

Using his food-training method, von Frisch found that bees not only possessed colour-sensitive eyes, but also keen 'noses', and were able to pick out one dilute floral scent among seven hundred others. He discovered that floral scents clung tenaciously to the bodies of returning workers, and their hive mates used the body odour as a clue to guide them to the food source. It was while von Frisch was experimenting with scent discrimination that he made a major discovery about how bees help each other to locate food. He noticed that when his bees had emptied one of the little nectar containers, they rapidly stopped visiting it. That was, in itself, not unusual. However, once a reconnoitring scout found the dish replenished with sugar solution, her sisters were buzzing around it within a minute or two. It dawned on von Frisch that bees must possess a superb intelligence system.

In the spring of 1919, von Frisch made an inspired observation which eventually led him to translate the amazing 'language' of honey bees. Working in Munich, in the attractive courtyard garden of the Zoological Institute, he sat watching bees sipping sugar water, and marked them with dabs of coloured paint. The bees soon finished the water, and took no more interest in the empty dish. He then refilled the container, and waited for a scout to alight on it. When one landed, he marked her, and carefully watched her behaviour when she returned to his special glass-sided observation hive. Later he wrote, 'I could scarcely believe my eyes. She performed a round dance on the honeycomb which greatly excited the marked foragers around her and caused them to fly back to the feeding place. This, I believe, was the most far-reaching observation of my life.'

Von Frisch was not the first person to have witnessed the 'dancing'. Several centuries previously, a monk who kept hives at Oberammergau had noticed the behaviour, and called it the 'ballet of the bees'. But von Frisch was the first to realise its significance as a method of communication between scouts and workers. Being a quiet, self-effacing man, he 'calmly smiled to himself', and, over a cup of tea, confided his discovery to a close colleague.

That summer saw the beginning of a long campaign of

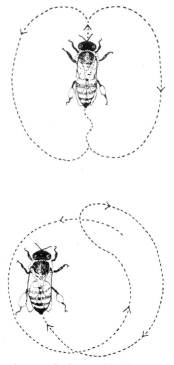

The waggle dance of the honey
bee takes the form of a figure of
eight (above). Below, the round
dance announces the find of
pollen and nectar within 50
metres of the hive

The dance of the bees. A
worker tells her companions
where to find nectar-rich
flowers by performing a waggle
dance

detective work to decipher the message in the honey bees'
recruiting dance. Initially, von Frisch thought that the circular
motion of the returning scouts simply stimulated their audience
to fly off at ever increasing distances from the hive until they
came across blooms bearing the scent that clung to their
informant. However, when he took his experimental food
dishes to locations more than 50 metres from the hive, the
pattern of dancing changed. Instead of performing 'round
dances', the marked workers altered their routine into a figure-
of-eight movement, vigorously shaking their tails during part
of it. Von Frisch christened it the 'waggle dance'. It suddenly
occurred to him that the dancing was not merely a general call
for help, but that the precise details of the distance and
direction of the food were encoded in the choreography.

Painstaking work over twenty years was needed to work
out the details of the honey bees' astonishing language. This
had to be carried out by watching bees dancing after they had
returned from feeding at known distances and directions from
the hive. Summer after summer, von Frisch recruited friends
and relatives to help him carry his hives around the countryside, sometimes suspending them from rocky peaks, and to
assist him to watch bees returning from dishes of sugar solution
placed in far-flung fields. Keeping track of scouts and workers
over distances of several kilometres called for methods which
were effective, if quaint. On one occasion, von Frisch had to
signal the departure of a marked scout from one of his feeding
stations by a blast on an old cow horn. This alerted his brother
Hans, some way away, who used a bell to stir a third helper
into action. He in turn sounded a trumpet to warn an observer
in the garden at Brunnwinkl to expect the arrival of the
returning bee, and to record its dance routine.

Von Frisch eventually realised that the round dance an-
nounced the find of a profitable source of pollen and nectar
within 50 metres of the hive. The richer the supply, the livelier
the dance. But it contained no directional clues at all. Food so
close could easily be located by the recruits. If the scouts
returned from over 50 metres, then the recruits needed better
guidance. They got this from the 'waggle dance'. It contained
information about both the distance and the direction of the
food supply. Another important part of this dance was the
'waggle run', in which the bee shook her 'tail'. The longer the
waggle run, the greater the distance. Von Frisch also discovered a relationship between the distance and the frequency
with which a scout danced. A bee returning from a food dish

placed 100 metres from the hive, danced ten times every fifteen seconds, whereas one which had flown two kilometres danced only five times during a similar period.

Von Frisch discovered that the recruits were told which direction to fly in by the orientation of the 'waggle run'. It could be seen in its simplest form when the weather was warm and sunny, and bees assembled outside the hive entrance. In these circumstances, workers, returning laden with pollen and nectar, frequently danced on the horizontal landing board. Here the 'waggle dance' was a rhythmic re-enactment of their outward flight, with the 'waggle run' pointing in the direction of the food, like the needle of a compass. Each bee kept her 'waggle run' heading in the true direction by taking a bearing off the sun. But once the worker entered the gloomy interior of the hive, she had to dance on the vertical comb face where her audience could no longer see the sky. She therefore solved the difficulty by converting the angle of her 'waggle run' with respect to the sun, into an angle with respect to gravity. If, for the sake of argument, the food-seeking flight should have been directly towards the sun, then her waggle run was performed vertically on the comb. If it was 30° to the right of the sun, then the waggle run would have been 30° to the right of the tug of gravity – and so on.

But supposing the sun was obscured? Von Frisch revealed that bees could overcome even that drawback. After the Second World War, he discovered that they could perceive the pattern of polarised light in blue sky. Provided that a bee could obtain a glimpse of it between the clouds, she was able to compute the sun's position, and so instinctively to calculate her course!

By unravelling perhaps the most complicated language of signs and signals in the animal kingdom, von Frisch undermined the view that lowly creatures – insects in particular – were deaf, dumb, and virtually blind. Of course, there were sceptics who repeatedly challenged his results because they were reluctant to believe that bees were able to communicate so effectively. But his discoveries have stood the test of time. Indeed, modern research has revealed that honey bees are even more remarkable than von Frisch imagined. For example, bees have formidable powers of learning, though they learn specific things only at certain times. One of Professor von Frisch's students, Elizabeth Opfinger, noticed in 1930 that a honey bee learns a flower's colour only when she flies towards the plant. Recent research by James Gould of Princeton University, New Jersey, has narrowed down the time this takes place to the final two seconds before landing. If she is tricked by experimentally changing the colour of the flower after she has landed, the new colour does not register in her mind. Once landed, the bee turns her attention to memorising the flower's perfume. Only when she flies away does she switch to taking in details of the landmarks which will help to guide her back after she has jettisoned her load of pollen and nectar in the hive.

Von Frisch's use of simple experiments, performed under conditions as natural as possible, influenced another group of zoologists who were trying to decipher the messages encapsulated in the colourful rituals of birds. Their studies formed a discipline to be called ethology – the comparative study of animal behaviour – and led to a great improvement in our understanding of courtship as communication.

It is not surprising that ethology had its roots in scientific bird-watching. Then, as now, one of the first lessons ornithologists learned was how to use behaviour as a guide to identification. In the field, feeding, flying, and courtship habits were often more useful than shape for recognising the species of a distant bird. And yet at the turn of the century, accurate accounts of bird behaviour hardly existed, and no one had tried to analyse even the simplest of courtship routines. This was to change. Zoologists with a liking for living animals started to make dossiers of habits which matched in precision the anatomical descriptions gleaned by their museum-bound colleagues from pickled specimens. One such person was Julian Huxley, the grandson of Darwin's bulldog, Thomas Henry Huxley. Julian Huxley believed that behaviour was as important as anatomy when evolutionary comparisons were to be made. He came to this view in the spring of 1912, when he spent an enjoyable holiday at Tring Reservoir watching the love ballet of great crested grebes. Both sexes were similar, sporting 'Elizabethan' ruffs – or tippets – and glossy black ear tufts. Huxley was enthralled by the elegant pas-de-deux ceremonies in which the courting couples performed almost every manoeuvre imaginable. They dived together and emerged in unison, their beaks full of weed; treading water, breast to breast, they twisted their heads ecstatically while showing off their tufts and tippets to their full splendour. Huxley made detailed accounts of several quite different displays – a 'penguin dance', a 'weed dance', and a 'cat display'. Being a fully trained professional zoologist as well as a bird-watcher, he also felt the need to divine the purpose of the various displays, and to propose how they might have evolved.

Three displays from the great crested grebe's courtship ritual. The 'cat display' (above), the spectacular 'weed dance' (middle), and the 'penguin dance' (below). After Huxley

Goose signals. A threat display (above), an alert posture, with neck upright (middle), and part of the beautiful courtship ritual (below). Right, a greylag gander (left) performs a 'triumph ceremony' to impress his mate

Huxley suggested that the mutual displays strengthened the grebes' pair bond, and established the emotional synchrony so necessary for successful mating to take place. He also noticed that a part of the grebes' ritual involved a rhythmic dipping of the beak into the back feathers. It appeared to be a feeble form of preening. Huxley called it 'habit preening', and so became the first person to realise that displays were fully or in part composed of common everyday actions, refashioned by natural selection to enhance their effectiveness as signals. He called this process 'ritualisation'. He envisaged that the original action, a preening movement, for example, might become more exaggerated and rhythmic, and that it would be made more thrilling to watch by the addition of colourful plumage, capes, plumes, tufts, or 'beards'.

While Huxley was extending his interest to the weird 'reptilian' courtship of red throated divers in Spitzbergen, Oskar Heinroth, the Director of the Aquarium at the Berlin Zoo, was comparing and contrasting the habits of closely related birds, and making progress in understanding their sign language. Heinroth would stride daily around the thickly wooded gardens before the public arrived, note-book at the ready, and record what he observed among the zoo's inmates. He was careful not to miss the little things that animals did, and this helped him to verify the usefulness of behaviour in assessing evolutionary relationships. For instance, he discovered that all members of the pigeon family shared the same technique of drinking; they all held their beaks in water, and pumped it into the gizzards by means of muscular contractions of the throat. Most other birds simply 'tip and sip'.

Heinroth's aim was to compile a dossier of behaviour for every kind of European bird, and he set about the task with characteristic vigour. He and his first wife Magdalena took no holidays, and avoided having children, the better to concentrate on their scientific work. Together, they raised most species from eggs so that they could watch their habits develop. It was an arduous task. Many of Heinroth's birds came to regard him as a parent; while doing his rounds, he was often accompanied by a retinue of imprinted wildfowl and cranes. Others sat contentedly beneath his office chair and resolutely defied all attempts to settle them on the zoo's ponds. By 1926, his inventories of behaviour were more or less complete. Between then and 1933, they were published in four illustrated volumes – *The Birds of Central Europe* – finally establishing that each kind of bird has its own unique behavioural repertoire.

Heinroth had a special love for ducks, geese, and swans, and paid particular attention to those of their habits which carried messages, and whose significance was understood by all of the members of a flock. He noticed that when a goose or swan was alarmed, it strained its neck upwards, perhaps to get a better view and as a preparation for take-off. Its companions immediately recognised this posture as a warning, and were put on guard. Also, when wildfowl met, they often went through the motions of sipping water, in much the same way that men drink together, not to slake their thirst but as a sign of friendship. By confining his attention to individual acts of behaviour which occurred over and over again in an unvarying manner, Heinroth was able to lay the foundation for a science of comparative behaviour which mirrored and complemented the discipline of comparative anatomy. He christened it 'ethology', using a term first employed in the English-speaking world by William Morton Wheeler, a pioneer in the study of insect societies. Wheeler was an American, who during the early years of the century sought a satisfactory word for the study of animal behaviour. Although Wheeler failed to establish the term, Heinroth was successful, and in the context in which he used it, ethology meant the 'comparative study of gesture'.

It was Konrad Lorenz who promoted Heinroth's new science, and took the analysis of wildfowl behaviour a step further. By comparing and contrasting their rituals, he was able not only to plot the relationships between species, but also to chart the way in which even the most puzzling displays had evolved.

Born in 1903, Lorenz was reared in Altenberg, a hamlet lying to the east of Vienna, between the Vienna Woods and the Danube. From an early age, he showed a propensity for keeping living creatures. As he grew up, so the animal population of his family home increased. There were crustaceans, fish, frogs, dogs, and birds of all kinds. At the age of six, he acquired some ducklings. Together with a little girl, Margarete, who was later to become his wife, he played 'ducks'. The hatchlings followed the children as though they were their real parents. Later, Lorenz reflected that early experiences often lead to a life-long endeavour. In his case, he was enchanted by the fluffy, piping ducklings, and became as imprinted on them as they clearly were on him! Certainly, ducks and geese became an abiding and deeply-rooted passion.

After graduating as a doctor, he rejected a medical career in order to devote his energies to studying behaviour under conditions as natural as possible. His skill as an ardent keeper of pets proved invaluable. Encouraged by Heinroth, Lorenz made Altenberg the centre for his research. He settled a colony of garrulous jackdaws in the roof of his father's pseudo-gothic mansion so that he could observe their social life at first hand. Broods of greylag geese had the freedom of the garden, and eventually the skeins winged their way back and forth between the river and the fields surrounding the house. While going about the village, Lorenz was invariably trailed by a gaggle of miscellaneous ducks and geese – a real-life Dr Dolittle – much to the amusement of his neighbours. But Lorenz took his ethology seriously.

Lorenz, being an excellent naturalist, knew that wild geese were downright difficult to observe. They were shy, inhabited uncomfortable swampy places, and panicked at the glint of a pair of binoculars. By rearing goslings from eggs collected from the Neusiedlersee, he obtained thoroughly tame birds of genuinely wild parentage which would unfold the details of their private lives as he sat among them. Between 1935 and 1938 Lorenz watched his greylags court, become betrothed and married; they mated, built nests, and reared their families under his endlessly inquiring eye. In the conduct of their daily life, they appeared to use formal signals which often involved the use of the head and neck in a kind of semaphore. In the spring, the young ganders searched for opportunities to impress their mates, threatening each other with low-held and outstretched necks, before victoriously rushing back to their partners to perform a 'triumph ceremony'. When attempting to persuade his goose to follow him into the water for mating, the gander held his neck in a graceful swan-like curve. When afloat, a lovely synchronous ceremony marked the exciting moments preceding copulation; both repeatedly dipped their heads, and sparkling drops of water tumbled down the pleated neck feathers, bouncing off the lovers' backs.

On the ponds around Altenberg, Lorenz watched drakes of all kinds displaying, each in its different way. A tame golden-eye flung its head back between its wings while kicking up a fountain of water behind its tail. Eiders bowed and cooed before their drab mates, while teal trilled and cocked their yellow vents at each other. At first sight, much of this behaviour looked confusing, the movements following each other with great rapidity. Lorenz finally managed to resolve each display into a series of distinct actions, and came to regard each as a signal which often served to show off some startling feature of the plumage. The golden-eye's head throw was given impact

The mandarin's display

The mandarin
drake in the mood
for courting

He begins by
ritual drinking

And follows this
by ritual wing
preening

His dramatic
head and wing
plumage are
thereby shown off

Like the shelduck (above), the mandarin's 'ritualised' courtship wing preening (below) has evolved from ordinary preening behaviour

by the white cheek spots; the teal's cocked tail flaunted the yellow vent feathers. These were what seemed to impress the ducks, and to arouse their urge to mate.

Lorenz was also able to take a little more of the mystery out of courtship. By comparing the 'signals' of closely-related species, he was able to make inspired guesses as to how some displays might have evolved. That of the gaudy mandarin drake was a case in point. When courting, this splendid bird erected his gorgeous head plumage and craned forward so that his beak just touched the water. Then, tossing his head over his shoulder, the mandarin ran his bill behind the bright orange 'sails' – modified wing feathers. The shelduck gave Lorenz a vital clue to the origins of the mandarin's baffling behaviour. During tense sexual encounters, the drakes often broke off their courtship routine and vigorously preened their wing feathers. It reminded Lorenz of someone scratching his head in frustration. It occurred to him that perhaps the mandarin's display was a highly 'ritualised' form of 'frustrated' wing preening. Lorenz's hunch seemed to be confirmed by the behaviour of the mallard, which showed a midway stage between the shelduck and mandarin. When confronted with a duck, a sexually-inclined mallard repeatedly sipped water and followed this up with a perfunctory pinion-preening gesture in which the brilliant metallic blue wing patch or 'speculum' was flashed in her eyes. Lorenz argued that the mandarin's courtship was essentially a similar routine, although the 'drinking' and 'preening' actions were reduced to symbolic formalities, and given dramatic impact by the evolution of exotic plumage.

Lorenz's strength lay in his quick eye and sympathetic understanding of animals. He was not a great experimenter, and proudly boasted that he never drew a graph in his life! It was a young Dutch zoologist, Nikolaas Tinbergen, who, through elegant experiments, revealed the precise nature of the signals to which animals reacted. In 1937, he joined Lorenz at Altenberg, and devised a series of tests to find out what signs caused young geese to take fright. While Lorenz sat with a group of newly-hatched greylags, Tinbergen hauled across the sky a series of flat models suspended on a wire 10 metres above the ground. He made discs, squares, and flight profiles of various kinds of birds. On the whole, the goslings took no notice of the models which soared above their heads. But when Tinbergen used the model falcon or hawk, the little birds reacted as though their lives were in peril. Goslings possess excellent vision, and it was inconceivable that they should confuse the roughly hewn models with living birds of prey. Tinbergen therefore set to work to discover what caused the goslings to be afraid. He constructed a very stylised model which when towed in one direction resembled a long-necked goose in flight; the goslings ignored it. When the model's motion was reversed, it became transformed into a long-tailed hawk; the baby geese piped in terror. It appeared as though young geese instinctively recognise birds of prey. We now know that the goslings were reacting to the shapes with which they were less familiar. Hawks are less numerous than wildfowl. The young geese rapidly get used to the sight of common birds, but retain a healthy suspicion of rare ones. Nevertheless, such simple experiments allowed Tinbergen to demonstrate that animals often react only to selective features of what they see – in this case, the proportion of the model in relation to its direction of motion. The recognition of key signs which trigger behaviour considerably advanced our understanding of how animal communication works. A happy combination of circumstances contributed towards this discovery.

When the dust of the Great War settled, and the grass started to grow back over the ravaged landscapes, people turned to the open air for pleasure and spiritual sustenance. Covered with polders and dissected with waterways, the flat Dutch countryside lent itself to the appreciation of natural beauty. Wherever the eye rested, wildfowl bobbed upon the waves, harriers quartered the reed beds, and gulls drifted like confetti across the wide skies. At weekends, and during school holidays, members of the Netherland Youth Federation for the Study of Nature (N.J.N.), could be seen cycling with their expert leaders along the dykes, on their way to watch birds in the coastal dunes, or to collect flowers among the Friesian cattle. Two teachers, E. Heimans and J. P. Thijsse, had written a best-selling Flora of Holland, and colourful natural history cards were given away in tins of Verkade biscuits and were avidly collected, like cigarette cards in England. This was the social milieu responsible for producing several very talented naturalists, among whom Niko Tinbergen was by far the most influential.

Tinbergen was exposed to wildlife from his early childhood. Like those of von Frisch, his parents rented a remote holiday cottage. This was located in central Holland, at Hulshorst, just below the Zuider Zee. It was a peaceful, attractive area of pine, birch, and beech forests, stretches of heathland, and dunes of

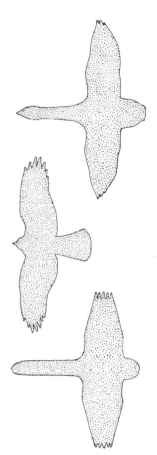

Selective features. The profile of a goose (above) and of a buzzard (middle). The stylised profile (below) is a harmless 'goose' if pulled to the left, but becomes a 'hawk' if pulled to the right, producing alarm in young goslings

silver sand. In the summer of 1929, just after completing his zoology finals at Leiden University, Tinbergen wandered among the heather of Hulshorst trying to decide upon a topic for his research thesis. Suddenly, he chanced upon a veritable town of hunting wasps – *Philanthus triangulum*. Each insect was about the size of an ordinary wasp, and had a bright orange-yellow tail. Intrigued, he watched them returning from their hunting expeditions to patches of bare sand, kicking the grains back to reveal their burrows. After taking their victims – honey bees – below for their grubs, the wasps surfaced and flew off in search of more prey. Here was a series of puzzles. With hundreds of wasps living fairly close together, how did each one manage to find its own home?

Jean Henri Fabre, the eminent French naturalist, thought that hunting wasps found their own burrows by the 'activation of a mysterious instinct'. Mysticism played no part in Tinbergen's approach. Inspired by the methods of von Frisch, he began his research with little more than a note-book, stool, and a pair of magnifying spectacles. But he was especially fortunate in possessing a knack for devising questions by altering small details of the animal's environment.

Tinbergen wondered whether the wasps were using landmarks. He noticed that before leaving on a foraging trip, the wasps quartered the ground in a manner which suggested that they were memorising the layout of the land. To test this theory, Tinbergen carefully changed the surrounding features while the wasp was away, removing twigs, pieces of grass, and pebbles from the vicinity of the burrow. When the wasp returned with a paralysed honey bee, she was seriously confused. Only after twenty minutes of trial and error did she accidentally find her nest entrance. Tinbergen then placed his own landmarks around the burrow: sixteen pine cones – pine cones were common on the ground where the wasp colony was located. When the wasp was happily going back and forth, he displaced the circle of cones a few inches to one side of the entrance. When the wasp returned, she flew straight to the centre of the ring, and so completely missed her nest. This and many other experiments demonstrated conclusively that the hunting wasps were taking their bearings from selective topographical details around their nest sites.

By 1935, the hunting wasps were declining fast on the heaths of northern Holland due to ever wetter and cooler summers. But there were plenty of other animals to investigate. Tinbergen's professor at Leiden asked him to prepare a six-

The stickleback's breeding signals. The bloated silver belly of a female stimulates the male to court her (above), whereas a red belly stimulates him to attack a rival (middle). But a red model is equally effective as a stimulus to aggression (below)

week practical course in behaviour for the students. This called for simple demonstrations. Tinbergen searched the ponds and ditches outside the town, and selected the common stickleback as a worthy demonstration-class animal. Although abundant throughout Holland, little was known about its habits. He and his colleagues therefore started to unravel the details of the stickleback's private life, using experimental techniques of the kind pioneered in the hunting wasp project.

Just before the outbreak of the Second World War, Tinbergen made a significant discovery about the nature of the stickleback's courtship signals. In the breeding season, the males became flashy little fish, with almost luminous turquoise eyes, green backs, and passionate red underparts. Among themselves, they became very pugnacious. Each would lay claim to a corner of an aquarium, and there build a small tunnel nest of algae into which he would try to entice the silvery females to lay their eggs. Tinbergen was keen to find out what caused a male relentlessly to attack another one, but persistently to court a member of the opposite sex. A casual observation helped him to solve the problem. At certain times of the day, the breeding males were driven frantic in their tanks which were resting upon the window ledges of the laboratory. He noticed that the excited behaviour coincided with the daily passage of the Dutch Royal Mail van. It was painted bright red, similar in colour to the belly of a sexually active male stickleback! This gave him an idea for some experiments.

Tinbergen made a number of model sticklebacks out of various kinds of coloured wax, introducing them one at a time on the end of wires into the territories of breeding males. Although the silver and grey ones were barely noticed, red models were viciously attacked no matter how crude they were. The same treatment was meted out even to a lump of red wax. On the other hand, when Tinbergen offered his male sticklebacks a crude model of a gravid female with a silver belly bloated with eggs, they courted it with as much enthusiasm as if it were a real fish. It therefore became apparent to Tinbergen that many of the details of a living fish were of no relevance in communication, and so were ignored. The male's behaviour was triggered by crucial and often very simple sign stimuli, or 'releasers'; a 'red' or 'bulbous silver' shape caused a male stickleback to attack or court respectively.

Further research by Tinbergen and many other people revealed 'releasers' to be a widespread feature of animal communication. For instance, David Lack, an Oxford zoologist,

Signal for aggression

A mounted specimen is placed in a robin's territory

The bright red breast feathers of the 'dummy' are a body signal which releases an immediate attack

The robin is driven to fury in its efforts to expel the 'intruder'

It continues to attack relentlessly

A herring gull chick begs by pecking at the red spot on its parent's beak (above). But a long, thin red object with white bands is even better as a stimulus

discovered that a bunch of red feathers placed in the territory of a robin was attacked in preference to a perfectly mounted effigy of a juvenile robin with a speckled brown breast. Fighting cock robins thrust their red chests towards each other. The fiery bib turned out to be a body signal which released an attack from a rival as surely as a key opened a door. The courtship antics of birds now also began to make sense. The avian body came to be regarded as a vehicle for rhythmically flashing signals for exciting and delighting hens or for repelling competitors.

One of Tinbergen's renowned pieces of research was on the signal which induced a herring gull chick to beg from its parent. The results also enabled him to improve on nature! Tinbergen noticed that a newly-hatched chick pecked upwards towards the tip of its parent's beak. The adult responded by regurgitating a mass of partially digested food, from which it would select a morsel, and gently offer it to the hungry chick. Several researchers had recorded that baby herring gulls tended to stab at anything red, even red cherries which hold no attraction as food for these flesh-eating birds. Tinbergen, who had been an inveterate gull watcher from his childhood, observed that the pecking of the chicks appeared to be aimed at the pronounced red spot on the lower mandible of the parent's otherwise yellow beak. By the simple technique of presenting day-old gull chicks with painted cardboard cut-outs of adult gull heads, the chicks themselves confirmed by their behaviour that the red spot was a releaser stimulating the begging. When Tinbergen gave them the choice, the hungry chicks preferred to peck a thoroughly misshapen dummy head sporting a red spot, rather than the head of a freshly dead adult on which the spot was concealed by a dab of yellow paint. Numerous tests revealed that, although the chicks showed an inborn preference for red spots, other colours were almost as acceptable, provided that there was plenty of contrast between the spot and the beak – a white spot on a dark grey beak was a perfectly good signal in the chicks' eyes. Once Tinbergen had discovered the features that caused a chick to beg, he was able to exaggerate them, creating a model 'beak' which oozed chick appeal. It took the form of a thin red rod, like a pencil, with three white bands close to the tip to enhance the 'contrast' quality of the 'releaser'. When offered to the chicks, they invariably pecked more compulsively towards the 'supernormal' image than to a high-fidelity gull's head.

Supernormal releasers were not simply invented by probing

scientists like Niko Tinbergen. A well-known one had been evolved quite naturally by nestling cuckoos to help them make willing slaves out of their foster parents. The colourful fleshy gapes of very young song birds operate as signs to encourage their parents to supply food, and to guide its delivery. On hatching, the cuckoo usurps the rightful brood, and overcomes the danger that the foster parents desert it by flashing at them a relatively huge and irresistible gape, into which they relentlessly deposit caterpillars.

After the war, ethology flourished under the leadership of people like Lorenz and Tinbergen. They and their fellow scientists focused their attention on what *caused* behaviour (e.g. signals which elicited reactions, like the red belly of the male stickleback), its *function* (e.g. to secure a private nesting territory), and how it *evolved*. By the late 1950s, the lens of ethological research had resolved an interesting picture of how signs and signals had been moulded by the process of evolution.

Tinbergen and his colleagues noticed that animals tended to display in situations where there was a difficulty of some kind or another, which were therefore fraught with tension. This appeared to be caused by conflicting emotions which competed with each other for expression. For instance, the arrival of a hen herring gull on the territory of a cock for which he had fought fearlessly, aroused mixed feelings. At first, he seemed to be seized by an impulse to drive her away – and he often did so – because she was an intruder. But if she behaved demurely, then he was, at the same time, sexually stimulated by her meekness. Ethologists suggested that opposing and incompatible impulses of this kind caused behavioural deadlocks which were broken in all sorts of ways, sometimes incongruously. Cock avocets – elegant piebald wading birds with tip-tilted bills – were seen to interrupt their courtship and to preen their breast feathers vigorously for several seconds before copulating; faced with a capricious hen, an amorous cock finch sometimes preened his wing (like Lorenz's shelduck), scratched his head or wiped his beak. From numerous observations like these, ethologists formulated the theory that conflict was one of the origins of sign languages. The strange and often irrelevant acts which animals performed when in an emotional turmoil became the raw material from which, over the ages, arresting displays evolved. Detailed analyses of the rituals even revealed emotional ingredients which could be separately measured – a pinch of aggression, a flavour of fear, a soupçon of sex, and so on. As more and more species were

A hen red cardinal feeding a goldfish. Fishes gape like young birds, so triggering the cardinal's parental instincts

An irresistible stimulus to feeding. A young cuckoo flashes its 'supernormal' gape at its foster parent – a reed warbler. Above, a reed warbler about to feed its brood. A young nestling's brightly coloured gape stimulates its parents to feed it

studied, it became apparent that sign languages were 'built' from ritualised forms of beak-wiping, nest-building, and movements signalling approach or avoidance, as well as preening. Even 'juvenile' behaviour appeared in the courtship of many birds: hen finches and gulls beg like babies, and their mates obligingly give them food.

Ethologists also discovered that the most spectacular display of any living creature – the peacock's – was based partly upon parental behaviour pressed into the service of sex. The peacock's vast tail is a sexual 'releaser' for charming peahens. The cock splays and sets it shimmering while strutting around, until a hen places herself strategically in front of him. He then bows stiffly so that the gorgeous fan arches over his head. If she is sufficiently beguiled by his 'ecstatic' posture, she crouches and invites him to mount her. Many kinds of pheasants possess displays which incorporate a bow or curtsey. In the red jungle fowl – the wild ancestor of the domestic chicken – the cockerel bends over and pecks at the ground in a ritual called 'tit-bitting'. This sends the hens wild with excitement, causing them to come running to his side where he can mate with them. Tit-bitting is similar to the method used by a mother hen to attract her brood of chicks to food. It therefore seems as though cockerels exploit the action to overcome the natural reluctance of the hens to approach too close to such a dominant and overbearing individual. During courtship, the resplendent Palawan peacock pheasant actually tips forward, picks up a morsel of food and offers it to the hen as a pre-mating delicacy. After she has accepted it, he spreads his wings and a tail sporting a magnificent array of glinting blue eye spots, and waltzes around her so that the light bounces off his metallic blue back and wings. In the peacock, tit-bitting is reduced to the bowing movement which precedes copulation.

The signs and signals an animal uses to communicate with its fellow creatures do not evolve in complete isolation, but are necessarily 'designed' to be effective against certain backgrounds – trees, open desert, or sea cliffs – without jeopardising the survival prospects of the signaller. Ethologists, with their method of studying ranges of closely related species, therefore addressed themselves to the question of how the environment shapes animal languages. Nowhere is the principle of 'signals for survival' better illustrated than in the behaviour of the kittiwake.

When Tinbergen moved to Oxford University in 1949, he encouraged many of his research students to become gull

Behavioural
deadlock

A cock avocet
wishes to mate
with the hen
(foreground), who
solicits copulation

But his drive to
mate may be held
in check by fear of
her

He preens in
frustration while
the hen continues
to solicit him

The crisis passes
and he eventually
mounts her

watchers. By means of straightforward field experiments, they attempted to assess the survival value of behaviour patterns. In this they met with a large measure of success. For instance, they demonstrated the advantages of having camouflaged eggs, and the crucial importance of removing shell fragments from the nest as soon as each chick hatched – the white inner surface of the fragments tended to draw the attention of crows to the vulnerable brood. In 1952, Tinbergen acquired as a student a post-graduate from Switzerland, Esther Cullen. By that time, a great deal had been learned about the habits of black-headed and herring gulls which bred on open sand dunes and flat coastal marshes. Tinbergen persuaded his new student to visit the Farne Islands to study kittiwakes to see whether they behaved differently from other gulls, and to determine whether the differences were related to their nesting on precipitous cliffs.

On a bright summer's day, nearly three centuries after the visit of John Ray and the Hon. Francis Willughby, Tinbergen and Cullen landed on Inner Farne. They sought evidence of natural selection as the force which shaped kittiwake behaviour to the demands of living on ledges. Esther was entranced by the lovely cliff-nesting gull, with its sloe-black eyes set in a flossy white head, lemon yellow beak, and pearl-grey wings, their tips jet black as though dipped in ink. For the next few years, the old medieval tower on the island was to be her spring and summer home, and the kittiwakes her waking obsession. She and her husband, Michael, who was interested in tern behaviour, were regularly visited and supervised by Tinbergen. While she carried out her dawn vigils in a wind-break erected perilously close to the edge of the cliff, he concentrated on capturing the kittiwakes' displays on film. What eventually emerged was a historically important account of how environment influences all aspects of an animal's be-haviour – including its language.

The first thing that struck Cullen was the kittiwakes' tame-ness. Unlike other gulls, which take fright easily and mob intruders mercilessly, kittiwakes sit calmly on their nests. Some even allowed her to touch them. She related this to the fact that their nests were plastered onto cramped ledges, and so virtually invulnerable to predators. Kittiwakes therefore lost the need to take flight and to harry robber gulls, which had great difficulty in alighting on the steep cliffs. And yet Cullen discovered that, among themselves, kittiwakes are extra-ordinarily quarrelsome when compared with herring gulls.

An adult kittiwake 'facing away' in appeasement (above). Below, the dark neck band of the chick, which is prominently displayed when it 'faces away'

Cliff-nesting kittiwakes. The nests are close together and safe from marauding predators. Breeding in such a situation strongly influences the birds' behaviour

Half-way to being a peacock

The courtship of the Palawan peacock pheasant

The male tips forward and picks up a juicy grub to present to the female

She accepts this pre-mating delicacy

The male then parades around her, showing off his colourful plumage

Right: The most magnificent sexual advertisement in the world – a male peacock displaying

If a herring gull finds itself without a nest site, it tends simply to settle on the outskirts of a colony. Such an option is not always available to a prospecting kittiwake. With suitable rocky shelves few and far between, it faces a permanent housing shortage. Living on ledges therefore accounts for the high level of territorial hostility, and the frequent use of the flame-red gape as a threat display to neighbours. All gulls courtship-feed, the cock regurgitating a morsel of fish or garbage onto the ground in front of the hen. The kittiwake presents his offering with more decorum, directly into his mate's gullet rather than onto the ledge from which it could so easily fall into the sea. Not surprisingly, the couples tend to sit facing the cliff, and the females mate sitting down rather than standing up like other gulls, so as not to topple over the edge.

Cullen discovered that kittiwakes also behave towards their offspring differently from other gulls. They do not possess a red 'bull's-eye' on the lower mandible for the chicks to aim at while begging; instead, the whole of the inside of the mouth is flaming red, and acts as a target for the brood. The food is taken directly from the parent's beak, not from the 'ground', presumably to prevent wastage over the side. The young chicks also have no need of the streaked brown camouflage of their more exposed cousins. They are therefore clothed in much paler gull-like plumage while still in the nest. The young kittiwake's greatest danger is being jostled into the sea by its parents arguing fractiously with neighbours, or by its nest-mates being over-zealous in competing for food. To safeguard its security, the kittiwake chick has a powerful appeasement signal to soothe whoever is causing the trouble. The gesture involves facing away from the troublesome parent or sibling to reveal the dark-brown collar to its fullest extent. All adult gulls employ a similar gesture, but in the case of the kittiwake the chicks too are able to use the signal. Under similar circumstances, other gull chicks could take to their heels and hide, but the young kittiwake has nowhere to go.

These and many other discoveries showed how effectively the kittiwake's behaviour had been tailored by evolution to meet the circumstances of living on ledges. Since then, numerous other ethologists have confirmed that the principle applies generally. For example, some kinds of signals are more suitable than others in certain kinds of country. Birds which skulk in dark thickets, or in dense reed beds where an ornithologist would lose sight of his outstretched hand, largely forsake visual signalling for calling. Sound carries around corners, so

Overleaf: Visual signals are crucial in open country. Ostriches possess plumes which, when flapped, can be seen from a great distance

Birds of deep cover often use their voice rather than their plumage for displaying. The yellow bittern's call carries a long way through dense reed beds

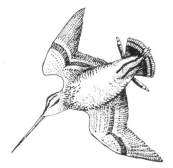

Prominent signal – a snipe drumming, its stiff outertail feathers bleating noisily as it dives through the air

species which inhabit such places tend to pour out a stream of song as a declaration of identity, to warn intruders, and to advertise for mates. It is therefore no accident that some of the world's most melodious songsters, such as babblers, bulbuls, and nightingales, are mostly drab brown birds adapted to deep cover.

Visual signalling comes into its own in open country. Bustards, cranes, and ostriches draw attention to themselves over long distances by flourishing plumes or by prancing around. Birds of prey which spend long periods gliding on thermals, carry identification marks on their wings, which bird-watchers also make use of as recognition signals. But an advertising bird makes itself conspicuous to predators as well as to mates and rivals, especially in exposed places. The problem has been solved in many species by a compromise: snipe and skylarks, for example, perform bold song flights, but, if they sense danger, they drop like stones to the ground, where their superbly camouflaged plumage helps them vanish from view. By contrast, those species which are less worried by natural enemies, such as poster-coloured parrots, peacocks, and bizarre Argus pheasants, can afford to display their finery with complete confidence.

When closely related varieties of birds share the same habitat, and are exposed to the danger of mixed and infertile unions, their sign languages tend to be very different. Although the females of many varieties of pheasant, birds of paradise, and of ducks are similar, their corresponding males are widely different, and often fabulously so, to deter cross breeding.

Research along lines pioneered by von Frisch, Lorenz and Tinbergen is still proceeding. For example, in Kenya's beautiful Amboseli National Park, lying among the northern foothills of Mount Kilimanjaro, Robert Seyfarth and his wife Dorothy have discovered that vervet monkeys possess an astounding repertoire of calls which appear to refer to different kinds of predators. To a casual onlooker, these black-faced monkeys squawk and chatter at random as they bound across the glades between the fever trees. But the Seyfarths noticed, over months of observation, that the presence of different predators caused the vervets to give voice to quite separate alarms, and that each alarm produced a different reaction among the members of the troop. Perhaps the monkeys had the makings of a primitive 'language'? The Seyfarths devised some tests designed to allow the monkeys themselves to demonstrate whether they 'understood' the meanings of their alarm calls. The American scientists carefully recorded the alarms and broadcast them through a loudspeaker placed in the bush. A pattern emerged. If they played back the alarm call signifying leopard, the monkeys often immediately fled to the outer branches of a tree, where these big cats cannot go. On the other hand, the sound of a martial eagle alarm call sent the animals diving for dense cover where they would be safe from swooping birds of prey. At the sound of a snake alarm, the monkeys were on their toes in an instant, peering into the grass for the offending predator.

Even with their language of alarms, vervet monkeys are taken frequently by all three predators – martial eagles, leopards, and pythons. The babies are especially prone to attack. It is therefore little wonder that when the Seyfarths suddenly revealed a stuffed leopard to one of their troops, the frightened animals went berserk, uttering their 'big cat' alarm from the tree tops for well over half an hour, although the leopard obviously did not move!

Further research in East Africa is revealing that a vervet monkey's amorphous grunts are uttered in a systematic fashion, and possibly passes all kinds of social messages. It may yet be proved that these and other primates possess a system of communication even more complex than von Frisch's honey bees!

LIVING TOGETHER

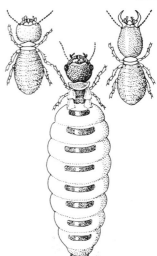

Insect society – castes of a typical termite colony. A worker (left), a soldier (right), and a primary queen (centre)

Ethologists have achieved much success in decoding the signs and signals which animals use to convey information to each other, enabling individuals to live in relative harmony together. The most recent advances in the discovery of animal behaviour have come from those zoologists who study the elaborate communities of monkeys, apes, and social insects. For the first time, they are able to provide a satisfactory explanation as to why animals should occasionally appear to help each other, and form organised societies.

Animals rarely live alone. When 'birds of a feather flock together', they become some of the most wonderful spectacles nature has to offer. A pinnacle of granite covered with tens of thousands of nesting gannets is every bit as exciting as Times Square, New York, or London's Piccadilly Circus. The eye is dazzled by the frenzy of birds, the smell of their guano is stifling, and their noise assaults the ear. But sociability is not confined to birds. At certain times of the year, terrestrial hermit crabs congregate to spawn on the beaches of Mona, a small Caribbean island. At first sight, the beaches appear to be carpeted with a mass of seething pebbles jostling for a position on the edge of the sea. On closer inspection, each 'pebble' is a snail-shell tenanted by an otherwise solitary hermit crab now seized by a desire for sexual companionship. During the southern summer, vast areas of the Antarctic seas are stained a vivid red by shoals of Euphausid shrimps or 'krill'; some of the swarms contain astronomical numbers of crustaceans, and are so dense that they form a soup, upon which the baleen whales feed. Anchovies, herring, and capelin also mass in great shoals, to the satisfaction of fishermen and seabirds alike. On land, our

Overleaf: A gannet colony, Grassholm, South Wales. Twenty thousand pairs nest a beak's length from one another. Such crowded breeding may stimulate the birds into laying their eggs at the same time

Alaskan fur seals, which breed in dense rookeries

crops are assailed by swarms of locusts and hoards of other insects, such as aphids and hoppers. Some mammals form huge herds, especially during migration. Wildebeest and caribou still make their annual treks in this manner, but in days gone by, bison, saiga antelope, and the extinct blaubok would migrate in herds which could stretch from horizon to horizon.

Superficially, the masses of crabs, krill, and caribou appear to be chaotic. But the confusion often conceals an underlying order. Flocks of swans and geese consist of separate families, each adult pair accompanied by their fledged offspring. But not all species organise themselves into human-like groups. Zebras, for example, are polygamous, each stallion keeping company with several mares and their foals; males without harems roam in bachelor parties. Among red deer, the sexes are segregated for much of the year; during the breeding season, the stags keep their distance from each other and try to tempt groups of hinds to join them for a while to mate. However, none of the higher animals can match the amazing social insects in the structure of their communities. Bees, ants, and termites maintain caste-ridden societies, with a sterile workforce and specialised reproductive males, dominated often by a single queen. They are capable of acts of apparent heroism, and will unstintingly give their lives in the service of their colonies, as anyone who has wandered too close to a beehive will bear witness.

Faced with such an array of different kinds of animal communities, the problem for zoologists has been to understand them. Even before Darwin and Wallace jointly proposed their theory of evolution to the Linnean Society in 1858, a few people accepted the idea that most animals were locked into a thoroughly selfish struggle for existence, with individuals vying with each other for personal advantage. But if success in competition was the very essence of survival, why should so many animals choose to live in crowds, where rivalry for food and other commodities must be enhanced. Cooperative behaviour was also difficult to explain. By helping another individual, an animal often puts itself at risk. The barbed sting of a honey bee, for example, is in use as lethal to its owner as to the unfortunate victim in which it is implanted. And there were still more puzzling questions to answer. Even Darwin was at a loss to explain how the sterile worker caste among social insects could have evolved. His theory demanded that advantageous characteristics – or behaviour – be handed on by breeding. How then could worker bees, ants, and termites pass on their favourable qualities to future generations if they themselves were unable to reproduce? The solution to these problems emerged only after a century of careful research into the social behaviour of animals.

After the publication of *On the Origin of Species*, scientists tended to search for evidence of competition and rivalry in nature. However, there were a few naturalists who were impressed by undeniable evidence of cooperation among animals. One of the first to be so was a French zoologist, A. V. Espinas. Through his study of animal societies, he was led to affirm in 1878 that no living being was truly solitary, because even the most steadfast 'loners' formed temporary attachments in order to mate. A few years later, Prince Kropotkin, a Russian anarchist, collated a mass of rather uncritical stories into a volume entitled *Mutual Aid*, in the vein of George Romanes, which purported to show that animals often assisted each other. In 1920 another attempt was made to explain social behaviour, when an American biologist, William Patten, saw it as part of a grand strategy of evolution. He theorised that life on earth may well have started as single cells, each making its own way in the world, as protozoans do today. He suggested that the next major advance took place when these isolated units joined forces, coalescing to form cooperatives of cells – or bodies. The next step came with the ordering of creatures into societies. Patten thought that this was a logical extension of a 'trend towards mutual aid'. In his scenario, the troop, tribe, or nest became the meaningful unit of biological accounting, not the individuals of which it was composed. This harked back to a very old idea that an ant colony or a wasps' nest behaved like a 'superorganism', its specialised caste system equivalent to organs.

Unfortunately, such views were scientifically worthless. They added nothing to knowledge, nor did they enhance understanding of why some animals should form intricate societies, and others not. As in so many areas of biological endeavour, it was Darwin who acted as a catalyst, provoking a number of useful investigations into social behaviour. The notion of evolution stimulated much interest in man's own origins, and this led many to investigate our closest relatives – monkeys and apes. As long ago as 1892, Richard Lynch Garner made a heroic, if outlandish, attempt to unravel the habits of wild apes.

Born in Virginia in 1848, Garner was destined to pursue a military career in the Third Tennessee Cavalry. But after a period of internment during the American Civil War, he shifted his ambitions from the army to apes – 'little effigies of the

Overleaf: By hiding among countless thousands of its fellows, an individual fish may be safer from predators

Garner and his jungle fortress

Army ants – a 'superorganism'. For many years, such a colony was considered to behave like a higher animal, with the different castes acting as tissues and organs

human race'. Initially it was their potential for speech that fascinated him, and he conducted some experiments along similar lines to those of von Osten and his horse, Clever Hans. They led nowhere. However, the study raised in Garner's mind the possibility of observing the great apes in their natural habitat, the equatorial rain forests of Africa.

On 9 July, 1892, he sailed from New York for the French Congo. In the depths of the jungle, he erected a stout square cage, constructed of steel and wire mesh with sides two metres long. He christened it Fort Gorilla. It was built, not to hold a bunch of powerful apes, but to contain himself, a young tame chimpanzee named Moses, and, from time to time, a native servant boy!

In those days, the tropical forest was generally feared. It was known to be infested with fever-carrying insects, poisonous serpents, and beasts with barbarous manners. Garner thought that it would be folly to ignore such real or imaginary dangers, and so the fortress was devised to withstand the onslaught of virtually anything, including a charging bull elephant. Inside Fort Gorilla, he maintained a civilised standard of living. Canvas curtains shielded him from the rain; he cooked on a kerosene stove, ate off a folding table, sat comfortably in a chair, and slept soundly on a bed. For his peace of mind, he kept a revolver, a rifle, and a heavy hunting knife within easy reach. For one hundred and twelve successive days and nights Garner kept vigil, and was rewarded for his enterprise.

Gorillas and chimpanzees wandered close to his strange jungle hermitage. On one occasion, a party of ten gorillas crossed the path that led to the fortified retreat. It was clear from what Garner saw, that gorillas were polygamous and nomadic, never spending the night twice in the same place. He also wrote of their 'incipient idea of government' with mature silver-backed males acting as leaders to whom all their wives and children deferred. On several occasions, Garner heard chimpanzees yelping and screaming as though they were having a 'carnival' around a fig tree laden with newly-ripened fruit. Although he filled his note-books with ideas that would not be considered scientific today, he had the correct approach to the study of animals, recording what they did in their native haunts.

The first genuine insight into primate and other animal societies was not obtained initially in the jungle, but in the farmyard, where Thorlief Schjelderup-Ebbe, a Norwegian, attempted to measure social relationships among chickens. In

doing so, he discovered the role of despotism as a force for order!

From childhood, Schjelderup-Ebbe had been absorbed in watching chicken behaviour. Just after the Great War, he applied in his research a technique just developed by Scandinavian naturalists, placing coloured rings or bands on the legs of his birds so that he could recognise each individual hen without difficulty. He then assembled flocks of colour-coded chickens and cockerels and simply counted the number of pecks each delivered to the other, carefully noting who pecked whom. His result revealed a rather surprising situation. Although chickens are highly companionable birds, always preferring to feed in flocks, their relationships are stratified by quarrelling. A hen's status is maintained by its pecking or threatening subordinates. When Schjelderup-Ebbe introduced several strange birds into his flock, a power-struggle immediately ensued. Soon a hierarchy was established in which the weaker or less forthright chickens deferred to the more aggressive ones. The outcome was a 'pecking-order' – an avian linc organisation of military precision. The bird which emerged at the top of the hierarchy pecked most frequently and with greatest confidence, and became the flock despot. Nevertheless, Schjelderup-Ebbe discovered that the brawling among his chickens was a transitory phase. Once the pecking order was established, all of the birds came to know their place in the flock. Thereafter, they rarely disputed their positions, although now and again there were revolts among the lowest ranks.

By switching his birds around, Schjelderup-Ebbe studied the personality traits that made for leadership or despotism. Cockerels were dominant over hens, but large hens did not necessarily dominate small ones. Chickens fighting in their own yard or run usually won more scraps than those fighting on unfamiliar territory – a phenomenon familiar in all human sports, teams and individuals tending to perform best on home ground.

The smooth operation of a dominance hierarchy depended upon the birds being able to recognise each other. Each chicken remembered who it could bully, and to whom it had to give way. This explained why the 'peck order' seemed to break down in very large assemblies of poultry, where there were too many faces for hard-pressed hens to register. Under these circumstances, order lapsed into anarchy. Straight-line hierarchies, of the kind that interested Schjelderup-Ebbe, occurred only in flocks of ten chickens, or less.

A peck order. Great tits establishing who has priority at the peanut basket

His observations went beyond poultry. The Norwegian zoologist studied a wide range of species, including house sparrows, pheasants, cockatoos, and canaries. What he saw convinced him that whenever two individuals meet, one will be subservient to the other. In fact, he became persuaded that he had discovered one of the major principles of existence – 'despotism'. He believed that everything had its despot. The storm was despot over the water, and water over the stone, which it dissolved. He even cited an old German proverb as evidence of the universality of the phenomenon: God is despot over the Devil!

Although Schjelderup-Ebbe took his idea too far, he nevertheless appreciated the fact that science thrives upon statistics, and showed his contemporaries how they could analyse groups of animals. Many scientists began to quantify behaviour, counting how often animals approached each other, and how often and for how long they groomed, fed, fought, and played. Many found the concept of rank order a useful one. It was a form of social spacing, not unlike territory. Each animal moved within its own personal space. Dominant individuals would not allow theirs to be violated, whereas low-ranking ones had little choice in the matter, and always had to retreat to keep their personal space intact.

The assertive behaviour which leads to the setting-up of peck orders is a product of evolution. It has obvious survival value to those at the top because they have the first choice of food and preferential access to mates. For example, during the spring, sage grouse gather to mate on traditional arenas on the North American prairies. Flaunting what at first sight appear to be a pair of poached eggs on its chest, and uttering a deep bubbling call, each cock vies with others to establish a place in the pecking order. The most aggressive and vigorous bird establishes himself in the centre of the arena. This is the area the majority of the hens will prefer; three-quarters of them will mate on his patch. But even the low-ranking cocks may be better off by accepting submission rather than opting for persistent quarrelling, during which they may be badly injured. At least, they stand some chance of breeding.

Schjelderup-Ebbe set a fashion for seeking dominance hierarchies in primate societies – some people are still searching. They have discovered, through counting and measuring behaviour, that baboons are especially rank conscious, with large males lording it over all and sundry. Sometimes there is one male in his prime who is clearly the undisputed overlord. The

Asserting mating rights

Two male black buck size each other up

They clash fiercely

After a struggle the weaker soon yields

The more powerful buck is left to mate

Solly Zuckerman in the 1930s

female baboons possess their own rank order. A female's position among her companions partly depends upon whether or not she has an infant. In baboon society, mothers are favoured, and when a female gives birth, she immediately becomes the focus of attention. However, high rank is no guarantee of indefinite and undisputed supremacy among savannah-dwelling baboons. Coalitions sometimes form among low-ranking animals, and these complicate the power structure. Seething discontent among subordinate male baboons can therefore build up into a revolt, toppling the overlords from their elevated social positions.

But animal societies, no less than human ones, cannot be understood solely in terms of rank orders. Measurements of status failed to explain the differences between one kind of organisation and another. They also failed to identify the nature of the social glue which bound otherwise fractious individuals together in uneasy companionship. It was Solly Zuckerman who discovered some of the forces of attraction at work in troops of primates, while analysing the hormonal basis of sexual behaviour.

Zuckerman first became acquainted with chacma baboons in South Africa, where he was born in 1904. At the time, people found little to admire in these animals. They were considered to be ugly, and offended standards of common decency by openly copulating. The South African government even subsidised the cost of campaigns to exterminate them. Zuckerman had no such aversion towards baboons. While studying at the clinical school of Cape Town University, his interest in baboons was aroused by the notion that they were relevant to an understanding of man's own roots. Most primates dwell in trees, but baboons, like ourselves, are terrestrial in habit. He thought that perhaps their way of life might throw some light onto the kind of society from which our immediate ancestors emerged. Zuckerman therefore decided to undertake a thorough investigation of the anatomy, growth, and behaviour of baboons. He addressed the task with enthusiasm, making periodic visits to a vast Cape farm near Graaf Reinet where several large troops roamed. Bedecked with field glasses, camera, and rifle, he often rode for miles on horse-back into the hills in search of baboons. Over a period, he mapped their home ranges and habits, and procured specimens. However, Zuckerman's ambition was to seek a living in London, the heart of the British Empire. This, he fulfilled in 1926. Qualifying in medicine two years later, but with no desire to practise his craft, he joined the staff of the London Zoo as a research anatomist and pathologist.

Between his professional duties as pathologist, Zuckerman found time to continue his primate work by keeping a close watch on a large colony of baboons on Monkey Hill – an edifice of reinforced concrete modelled to resemble a cliff face. The inhabitants of the hill were of a different kind from those which lived in South Africa. They were hamadryas baboons – the sacred baboon of Egypt – swarthy animals with rather stocky legs, admirably suited for scrambling around steep rocky gorges. Although the females were undistinguished-looking creatures, their overlords sported dog-like faces and bare buttocks in matching pink. Their visages were framed in tufts of hair, and impressive capes of flowing grey fur covered their shoulders. Sadly, life for many males on Monkey Hill tended to be short and savage. In those days, the management of animals in zoos fell far short of being scientific. With a flair for knowing what would set the turnstiles clicking, the curator had decided to populate the enclosure with showy male baboons. When the hill was opened in 1925, one hundred baboons were released, all but six being males. With such a heavily-distorted sex ratio, it became an eventful, if macabre, exhibit. The 'accidental' introduction of those six females led to wholesale carnage, as the rival males violently fought each other for their sexual rights. Twenty-seven animals were killed in the fracas. When Zuckerman started to monitor Monkey Hill, the sex ratio had been marginally improved, the troop being augmented by a further batch of thirty female baboons. Nevertheless, they died often enough, enabling him to examine the state of the ovaries and testes. He discovered that the waxing and waning of the female's sexual skin – the naked and highly-coloured area at the base of the tail – varied with the ovarian cycle, reaching a peak at ovulation when she was most fertile and sought after by the males. Such research inevitably led Zuckerman to consider the behaviour which may have been under the control of sex hormones.

He discovered that the social life of the baboons was built upon the principle of dominance, but this was overridden by powerful attractive forces based upon sex. Although the males did strike up friendships with each other, Zuckerman was left in no doubt that each overlord attempted to shepherd as many females as he could manage. Under the special circumstances of Monkey Hill, that was not many! The males were always potent, and always approached any female that showed signs

Overleaf: A male hamadryas baboon and his harem of females, together with their young

of being sexually receptive. With hamadryas baboons, as with all monkeys and chimpanzees, there was never any mistake about when this happened because the female's sexual skin bloomed and deflated in rhythm with her desire to mate. When tumescent and blushing, it was an open invitation for males to mate. But even when they were outside the fertile period of their menstrual cycle, female baboons often consented to copulate. Zuckerman realised that much of the mating he witnessed in the zoo had nothing to do with procreation, but had a social purpose. In many ways, the manners on Monkey Hill resembled those of a bordello. Females were able to obtain food or approach a despotic overlord, escaping punishment simply by proffering their hind quarters. Sex therefore allowed them to reap rewards to which their status did not entitle them.

Zuckerman believed that baboons were also kept together by their mutual fascination with fur. Babies spent their early life clinging to, and taking comfort in, their mother's pellage. As they grew up, their attraction to fur expressed itself in long hours spent grooming friends and mates. Harry Harlow, the experimental psychologist from Wisconsin, rediscovered the importance of fur some twenty years later, and called the phenomenon 'contact comfort'. Nevertheless, Zuckerman felt that a permanent interest in sex largely accounted for the cohesiveness of baboon and other primate societies. Each over-lord kept his females close by to satisfy his smouldering sexual appetite. The females – he suggested – while submitting to his demands, kept him company for fear of being thrashed by him should they wander too far from his side. The comparison with man's own sexuality was obvious; regular love-making, far beyond the requirements for producing children, led to a strengthening of bonds between men and women.

Later field investigations confirmed that Zuckerman was influenced by the exaggerated amount of mating and fighting among the baboons of Monkey Hill. This was caused by a sex ratio so distorted in favour of males, and by the overcrowding. But he was one of the first scientists to record accurately what he saw taking place inside an animal community. However, it was necessary to watch primates in their natural surroundings to confirm that there is more to monkey society than sex.

In the 1930s, watching wild animals was considered to be a mildly eccentric form of behaviour for respectable people. There was little financial support for those who wished to go 'native'. Those who wanted to study primates found it easier to raise funds than most, because they were able to argue that the study of monkeys and apes would reveal truths about man's own evolution. One such was Clarence Ray Carpenter, a young American psychologist. On Christmas Day, 1931, he began his observations of black howler monkeys on Barro Colorado, a small forested island formed by the damming of the Chagres River during the construction of the Panama Canal. Unlike Garner, Carpenter was not apprehensive about the jungle. Tracking his quarry on foot, he spent days craning upwards to watch howlers through binoculars as they moved around the tops of trees thirty metres high. Like Schjelderup-Ebbe and Zuckerman, he systematically quantified the howlers' activities, counting the number of males, females, and immatures, noting where they went, what they fed on, and so on.

He discovered that, on Barro Colorado, the howlers lived in stable groups, each composed on average of three adult males, seven females, four juveniles, and three infants. Each group defended an exclusive home range against trespassing neighbours. The roaring for which the howlers were renowned was considered by Carpenter to be a form of choral conflict. (Since then, the vocal behaviour has been reinterpreted: the groups locate each other by their bellowing, enabling them to adjust their movements to keep apart.) There was little sign of the rampant sexuality witnessed by Zuckerman in London Zoo's baboons.

In 1937, Carpenter, then a Research Associate of the Peabody Museum at Harvard, took part in the Asiatic Primate Expedition (A.P.E.). His objective was to make a thorough study of wild gibbons in their undisturbed natural environment. The gibbon was of special interest. It was the smallest, and despite being the most numerous, the least known of the apes. Carpenter's project was also spurred on by the familiar idea that much can be learnt about man by observing the habits of primates. Humans, like gibbons, possess comparatively long arms, and this led to the theory that human ancestors might have swung through the forest as they do. Furthermore, the gibbon's social life may have suggested a pattern on which our own ancestral society was based. This weighty expedition, which included three other primate specialists – Sherwood Washburn, Adolf Schultz, and Harold Coolidge – was based in the mountainous area of northern Siam, now Thailand. Carpenter's camp was established on the lower slopes of Doi Angka, the highest mountain in Siam. The area was covered in virgin rain forest and criss-crossed by deep valleys which resonated to the wailing of gibbons.

Carpenter was by now a highly accomplished 'monkey watcher'. In the bosom of the jungle, though, the atmosphere was humid, and the heat sapped the power of his muscles. The massive trees loomed upwards for thirty or forty metres before their canopies arched over, filtering the sunlight into a diffuse green light. On the forest floor, thorny shrubs, tangled lianas, and great sharp leaf blades tore at clothes and flesh alike, and snatched at his legs. Ticks and leeches were ubiquitous. It was a strangely still world where little seemed to happen. Unseen birds occasionally delivered their staccato songs from the dark recesses of the vegetation. Now and again a barking deer called in alarm. Here and there fat but harmless forest scorpions scurried underfoot. The canopy, where the sunlight generated huge amounts of foliage, flowers, and nourishing fruit, was the most hospitable environment for wildlife, and that is where the gibbons lived.

Carpenter followed twenty-one separate groups of acrobatic and energetic gibbons as they brachiated with uncanny ease, gracefully launching themselves across gaps twelve metres wide. Quite often, they walked upright along the larger branches, using their arms for balance; *Hylobates*, their scientific name, means 'tree traveller'. Carpenter made hundreds of pages of notes, and supplemented his observations with ciné film and sound recordings. In those days, recording in such remote areas was a troublesome undertaking, involving the use of battery-powered disc-cutting equipment. This was linked to a vast parabolic reflector two metres wide. The cumbersome 'semi-portable' contraption worked so well in focusing the sound of distant gibbons, that they often seemed to be hollering right in front of Carpenter's microphone! He found that the best time to record was about eight o'clock in the morning, just as the rays of the rising sun 'cheered the hearts of the gibbons', which then sang joyously.

Carpenter discovered that these small arboreal apes were the 'birds' of the primate world, and very different from howler monkeys and baboons. Not only did they sing, but, like many birds, they moved around in monogamous family groups. Accompanying the adult pair were up to four of their immature offspring, the eldest of which was sometimes as much as eight years old. The babies were nursed for nearly ten months and, while small, would cling to their mother's belly while she took terrifying leaps along her tree-top trails. Carpenter watched the younger members of the families playing tag – as human children do.

Siamangs, which like gibbons live in family groups in the rain forests of the Far East

The importance to the families of spacing was obvious to Carpenter. Each one had its own part of the forest which was defended by bouts of chorusing and, if necessary, through violent battles. The business was so crucial that ten per cent of the male's waking hours were devoted to territorial proclamation, either by singing or through spectacular displays of tree-top bravado before an audience of his neighbours. There was no let-up, and the regular early-morning gibbon symphonies probably served the families as a daily re-establishment of ownership.

Although he searched for evidence of dominance hierarchies, Carpenter found none. The male and female gibbons appeared to be of equal status. Nevertheless, a gibbon was clearly dominant over all comers on its home ground. If much of the gibbon's relationships with neighbours was volatile and aggressive, those within the family group were often tranquil. Carpenter observed them spending long periods at their toilet, cleaning each other's fur. A gibbon being groomed gave the appearance of loving every moment of the experience, languidly draping itself across a branch, constantly shifting position and arranging its loose limbs this way and that so as to provide its partner with access to every part of the body. Suddenly the roles would change, the groomer becoming the groomee. Then, as now, those who have studied primates have come to the inescapable conclusion that the time these animals spend servicing each other's coats far exceeds the demands of cleanliness. The behaviour probably has a vital social purpose, cementing friendships and reinforcing relationships. The desire to finger another's fur and to be groomed may play just as important a role in monkey and ape societies as the sexual attraction suggested by Solly Zuckerman.

Carpenter's detailed observations of gibbons in Siam altered the status of these lovely animals from being one of the least known to the most thoroughly studied of the apes. It transpired that, as with most primates, they lived in structured groups which do not alter from one part of the year to another. This state of affairs contrasts markedly with many varieties of animal whose social life is nothing short of turbulent, altering with the seasons. One such animal is Hooker's sea lion.

Sea lions spend most of the year at sea pursuing fish, with little inclination for each other's company. However, during the southern spring, their character changes. The massive bulls are first to react to the urge to reproduce. They haul themselves out onto their traditional breeding beaches on Enderby Island,

Large males have evolved through competition for females. The size and power of this bull Hooker's sea lion enables him to secure many females

a bleak speck in the ocean south of New Zealand. They do not tolerate each other's presence; by threatening and sparring, they become spaced out and masters of their own patches of sand. During December and the first half of January the cows emerge from the surf until the shore supports a busy, dense crowd of about one thousand animals. The cows have placid natures, and are content in each other's company. They tend to loaf around areas of their own choosing, the bulls having little control over where they finally settle. However, each bull's prime concern is to occupy a territory favoured by the cows. At first, the bulls are fairly calm. But when they are faced with the daily arrival of ever-increasing numbers of potential breeding partners, they become excited and bad tempered. Border disputes become more frequent as the bulls press their territorial claims ever closer to the spots occupied by the cows. These often lead to mass brawls with as many as thirty or forty bulls edging in on the fighting pair, struggling to stake a claim to a choice spot. So intense is the competition during the peak period, that some bulls may maintain a presence in areas frequented by cows for only a few minutes, while the biggest and most belligerent ones may reign supreme as beachmasters for a week or more.

Once a cow has run the gauntlet of bulls patrolling immediately seaward of the surf, she dashes up the beach, avoiding the attentions of the territory owners, until she reaches a bevy of resting females. She is not ready for sex until she has given birth to a pup. About one week after the event, she goes around soliciting mating. There is little finesse in the courtship of Hooker's sea lion. When in heat, the female's sexual perfume fires the sensibilities of the bulls. Lying prone, she takes one of the hefty creatures on her back. For fifteen minutes or so, they mate. But then her protests bring the business to an end. She will not indulge in further copulation for another year.

Meanwhile, the pups assemble in groups or 'pods' of up to fifty, away from the centre of bull activity. Like most mammals, but unlike Carpenter's gibbons, the males take no interest in the offspring, which were fathered by the previous year's beachmasters. In fact, the violent behaviour of the bulls often results in baby seals being trampled to death. A pup's only protector is its mother, who, recognising its smell and voice, allows it to suckle. The pods are rather similar to play groups, with the young animals indulging in mock fighting and chasing games.

At the end of February, the structured sea-lion society

fragments. The pods break up, the cows and their offspring vanishing out to sea. The exhausted beachmasters no longer possess the territorial imperative; their passion is spent. As daylight dwindles, the sea lions once more become recluses, and all is quiet on Enderby Island.

There are many kinds of animals whose social structures alter over the course of a year. For instance, most seabirds and the majority of hoofed mammals re-order themselves for breeding. One of the first naturalists to describe such structural alterations was Frank Fraser Darling. He surveyed populations of grey seals and red deer in a more detailed way than had been achieved before for any species. His work set a very high standard for scientific naturalists for decades afterwards.

Born in Yorkshire in 1903, he was a tall gently spoken man who developed a great love for the Highlands and Islands of Scotland, and their wildlife. Unlike many of his colleagues who considered animal behaviour to be something that should be studied in the laboratory, Fraser Darling was keen to get his boots dirty and to learn how animals survive in the wild. He had difficulty raising finance, but eventually succeeded, and between 1934 and 1936 he made what is now considered to be a classic investigation of red deer in Wester Ross.

His methods were considered to be unusual at the time. For months, he lived with red deer, camping in their countryside and getting to know them as individuals. This enabled him to 'soak up the environment', and to think like his animals – always the hallmark of a good naturalist or hunter. The deer were so shy that he had camouflaged tunics specially made; even his handkerchiefs were khaki-coloured. His tent was of olive-green fabric so that it blended into the Scottish landscape. From dawn until dusk, he stalked deer through the glens, often bare-footed, observing his quarry through binoculars and telescope. He became literally a scientific hunter, pursuing the deer until he could predict their movements, and could understand their problems, such as braving the weather and the flies. He was forever recording what they were doing.

When his work on red deer was completed, he turned to seabirds, and later, between August and November 1937, he lived on the Treshnish Isles, off the west coast of Mull, where, with the help of his first wife, Marian (Bobbie), he watched grey seals. (This was a trial run for a seal project on North Rona, a wild and remote island, from which he eventually had to be rescued by the Royal Navy during the Munich crisis.) On Treshnish, he cut a fine figure in his Hunting Fraser kilt and

Overleaf: Lethal prowlers. By hunting in a pack, hyenas are able to outmanoeuvre and overpower even animals as large as zebra stallions

Harris tweed jacket, both unceremoniously shed whenever he needed to stalk close to resting seals. On the occasion of his wedding anniversary in 1937, he managed to get through the 'barrier of a bull's watchfulness', ending up lying alongside it. He then had the problem of crawling back through the skerries to the shore, without disturbing the dozing animal.

Despite appalling weather – his camp was lashed by gales, and drenched with rain – he mapped the territories of the bulls, and discovered where the cows dropped their calves, watching the way the mothers used their ample bodies to shield their white-coated babies from the pounding breakers. It seemed to him that the big dominant bulls served the majority of the cows. He interpreted this as 'good for the species'. Likewise, the forceful and most aggressive stags held the lusher pastures in the glens, which attracted the hinds; those stags therefore mated most. He argued that they were the best or 'fittest' of their kind, and by fathering most of the fawns, kept the deer stock strong.

Extended field expeditions gave Fraser Darling plenty of time to contemplate the value of animals living together. He concluded that when there were predators about, the more eyes and ears, the better. Not surprisingly, prey species often tended towards some kind of gregariousness. In the case of small birds (finches and waders), the sight of a peregrine falcon wheeling overhead causes them to close ranks, presenting the enemy with a tightly bunched flock into which it would be difficult, or even suicidal, to stoop. By contrast predators tend to be solitary hunters, like leopards and leopard seals, though some operate in teams with consequent advantages. A pack of Cape hunting dogs can bring a full-grown zebra to its knees, something that is well beyond the capability of a single animal. A host of birds feed all the better in flocks; starlings, for example, act as beaters for each other, startling insects and bringing them to the attention of hungry companions. Cormorants, when fishing in flocks, drive their prey before them. The frenzy of crowded fish makes it easy for the birds to seize their quarry as soon as they dip their beaks into the water. This dramatic behaviour takes place in the beautiful Indian sanctuary of Bharatpur, where vast rafts of cormorants and shags frequently assemble, preying on dense shoals of fish. Mate selection may also be helped by social gatherings. This may be one of the advantages of colonial nesting for sea birds. The crowded colonies themselves may act as erotic theatres, with the huge numbers of displaying creatures helping to raise the

Cooperative fishing by white pelicans

level of excitement, thus assisting synchronous nesting. This benefits the prey species, because the predators cannot possibly devour the sudden glut of eggs and chicks. Their sheer numbers swamp the 'market'.

The Second World War interrupted the collection of information about animal societies. But once life returned to normal, scientists emulating Carpenter and Fraser Darling went out into the wildernesses to make in-depth social biographies. East Africa, with her extravagant riches of big game, lured many of them, including the German zoologist, Bernard Grzimek. They were eager to camp in the bush, sometimes shadowing their animals on foot, sometimes from vehicles. Many monitored their quarry from light aircraft. Getting stuck in hyena holes became a routine occurrence; iron-hard termite mounds projecting from inadequate air strips were a regular hazard. Through tough living and sheer perseverance, this new breed of zoologist discovered patterns of behaviour and webs of relationships hitherto unsuspected, as a few examples will illustrate.

The East African antelopes revealed their secrets to a handful of biologists. The Uganda kob proved to possess a most interesting breeding system. H. Buechner found out that the males arranged themselves into a communal arena for breeding. Within the traditional mating grounds, each buck strove for his own territory, about the size of a putting green, where he could offer the does a spread of grass like a well-laid dining table. Up to forty males defended their little pieces of real estate on the breeding grounds, waiting for the does to come and copulate. Like the 'lek' of sage grouse, the appeal of a buck's territory depended upon its position within the display ground. It was the central ones that were most hotly contested by the males, and to which most of the females gravitated in order to mate.

Hans Kruuk, a former student of Niko Tinbergen, struck upon the fact that the hyenas in the Ngorongoro Crater, Tanzania, had turned the tables on lions. The clans behaved like packs of wolves. Far from feeding as scavengers, they were formidable social hunters, from whose kills the King of Beasts fought for a scrap or two. Nevertheless, on the Serengeti Plains, hyenas tended to fulfil their more traditional role as nature's refuse-disposal agents. Clearly, even within one species, behaviour could differ from one place to another. Meanwhile, in the Budongo Forest, Uganda, Vernon and Frankie Reynolds were discovering that the chimpanzee had a curious social life

The communal mating ground of the Uganda kob. Each buck has its own patch, into which it will try to attract the does

of shifting relationships which appeared to depend upon the relative abundance of food. When figs were scarce, the apes were quiet and slunk around the jungle often alone. But when fruit was plentiful, the chimps were boisterous, noisy, and sociable. Another group of chimpanzees was studied by a Dutchman, Adriaan Kortlandt, and they demonstrated to him how they could gang up and attack a 'stuffed' leopard with sticks and stones.

During the latter part of the 1960s, George Schaller, an American, made a reputation for himself as an intrepid and resourceful researcher. After analysing the social behaviour of musk ox on Baffin Island, he tackled the mountain gorillas which lived on the forested slopes of the Virunga Mountains, straddling the borders of Uganda, Rwanda, and Zaire. By tracking them discreetly on foot, he managed what everyone thought to be impossible; he gained the confidence of the great apes. They even allowed him to sit on the edge of their groups. Far from being dangerous – at least when not threatened – he found them docile, living in extended families, and surviving on a diet of wild celery and goose grass. Afterwards, he switched his attention to the tigers of Khana, in central India, following them on foot!

One of Adriaan Kortlandt's chimpanzees attacking a stuffed leopard with a stick

The renowned palaeontologist, Louis Leakey, encouraged two women to take an interest in the African apes. Today, they are well known, through appearing in films sponsored by the National Geographic Society of America. Dian Fossey, a physiotherapist, continued Schaller's observation of mountain gorillas, and Jane Goodall made friends with the chimpanzees at Gombe on the shores of Lake Tanganyika. She discovered that chimpanzees regularly used sticks to 'fish' for termites, and would kill smaller monkeys for their flesh.

Through the accumulation of knowledge, it became increasingly apparent that social behaviour and the pattern of animal communities were strongly influenced by the environment. For example, different sorts of colobus monkeys possess different kinds of life style, and this is related to their individual feeding requirements. Both the red, and black and white colobus live in the Kibale Forest, Uganda, but the distribution of their food in time and space differs. The red colobus wander through the tree-tops in troops of up to fifty or so individuals. These monkeys exist on fruits, flowers, and flushes of young leaves. Since the periods when each of these are available are staggered among the various varieties of trees, red colobus need large home ranges in order to supply their dietary needs.

Gentle giants – a troop of mountain gorillas at home in the highlands of Rwanda

By contrast, the pretty black and white colobus has a compli-cated stomach which enables it to extract nourishment from mature leaves. With such a potent digestive system, all that one of these monkeys has to do is to stretch out its hand, and it will find a fistful of food. It therefore has small territorial requirements because just a few trees will furnish sufficient foliage to support a group.

Predation pressure and competition for breeding, especially among males, also leave their stamps on the way animals organise themselves. Life may be safer for a monkey living in the tree-tops than for its relatives which seek their livelihood on the ground. Perhaps it is no accident that the relatively terrestrial baboons and macaques tend to form large troops, with bold, hot-tempered males capable of ganging up against a common enemy. But the race to mate may also lead to the evolution of impressively muscular males. They can expect to win conflicts with punier rivals, and so sustain the best breed-ing record among their kind. This accounts for the huge sexual differences in size in such animals as red deer and Hooker's sea lion. However, among the primates, with their well-developed societies, this phenomenon raised the question of why the dominant mountain gorilla or baboon should tolerate other males in his troop? Why does not he expel them, establishing for himself a complete monopoly of mating rights, unchallenged by subordinate rivals? But this puzzle was part of a wider problem to which zoologists began to address themselves. How could social behaviour evolve when the evolutionary process favoured the most thrusting, selfish individual, pursuing its own personal advantage in the struggle to survive?

The crux of the problem lies in the fact that sociability implies a degree of cooperation between creatures which keep company. Cooperative behaviour used to be explained away as 'good for the species'. Although a pleasant sentiment, natural selection, the engine of evolution, does not recognise acts which are 'good' for the flock, herd, or species. The goldfinch which pipes up in alarm, warning its companions of a swooping sparrowhawk, may capture the predator's attention by so calling, and be killed. If it left no offspring, its noble behaviour would become stamped out. According to the arguments initi-ated by Darwin, the bird which shrieks a warning *must* some-how be acting in a manner beneficial to itself. The cry may cause the flock of finches to scatter, confusing the hawk, and making it less likely to kill any bird, including the caller. Another possibility is that, although the individual raising the alarm may be slightly disadvantaged, it may be considerably worse off by keeping quiet. The hawk may then pick one of the finches at random, possibly the alert, but silent, bird. What, on the surface, seems to be an act of selfless heroism may turn out to be one of considerable self-interest.

A whole spectrum of social behaviour in insects is less easy to explain, especially where breeding is the prerogative of a minority. Acts of suicide by sterile workers in defence of their nests are commonplace. Two decades ago, few zoologists recognised that here was a problem which gnawed at the very roots of Darwinism. But one naturalist did so, and eventually came up with a solution.

From an early age, William Hamilton suffered from the stings of honey bees in his parents' garden in Kent. He nevertheless developed a keen interest in insects, especially butterflies and moths, which he watched and collected on the chalk downs behind his home. A few gaudily coloured specimens particularly engaged his attention. They were often distasteful, if not downright poisonous, to birds, and their bright markings were designed to warn predators to leave them alone. In one species, the cinnabar moth, Hamilton was puzzled by the details of the warning process. The caterpillars looked evil; they were striped black and yellow, and disported themselves brazenly in groups on the saffron flowers of ragwort. When a naive bird spied them, it would seize a hapless grub in its beak, immediately rejecting it because of its horrible flavour. Thereafter, the cinnabar caterpillars' uniform was indelibly stamped onto the bird's memory, and it studiously refrained from attempting to eat any other creature bearing similar coloration. Hamilton saw that the moth species benefited, but how did the sacrificial victim profit? He came to the conclusion that the sacrifice was probably worth making because the surviving cinnabar caterpillars on the same plant were probably hatched from a single batch of eggs. In other words, the dead victim and the lucky survivors were close *relations*.

Kinship was the crucial factor in understanding social behaviour. Hamilton was not the first person to experience this flash of insight. Thirty years before him, the eminent geneticist R. A. Fisher realised that the degree of relatedness determined the lengths to which animals and people might compete or cooperate among themselves. He buried the germ of this idea in a highly technical book, *The Genetical Theory of Natural Selection*. At university, Hamilton devoured and digested the contents, impressed by the potential of mathematics as a powerful

Overleaf: A female wildebeest endangering herself by defending her calf. Such parental behaviour is an example of a general phenomenon of 'aiding relatives'

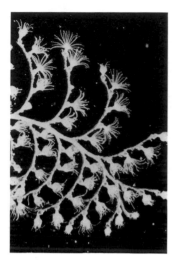

tool for clarifying and resolving complex biological problems. He became obsessed by the implications of kinship as a force for shaping social behaviour. After graduating, he registered for a Ph.D. at Imperial College, London, becoming a recluse, wrestling with algebraic equations in libraries, on park benches, and in his mean 'digs' in Ealing. Many original thoughts came to him while he was riding on London buses, or resorting to late-night contemplation sessions on Waterloo Station!

The essence of his thesis was this: although we all tend to think that evolution is nudged along by the process of 'survival of the fittest', in the long term, no creature, not even the 'fittest', escapes death. What carries on are copies of its genes – the atoms of inheritance. Through acts of reproduction, these are replicated, shuffled with those of sexual partners, and bequeathed to successive generations. Genes therefore have a kind of immortality. Natural selection singles out those genes which contribute to their own survival, regardless of which individual they reside in. It therefore matters nothing that an animal jeopardises its own life or breeding potential, providing that by doing so, it promotes the survival of copies of its own genes carried in others. The simplest case concerns the help parents give to their offspring, which inherit a sizeable proportion of their own genes. Hamilton's clever calculus revealed that parental behaviour was a special example of a general phenomenon of 'aid given to relatives'. If evolution could favour cooperative behaviour by parents towards their children, then it could also promote, to varying degrees, help offered to brothers, sisters, and more distant relatives. After all, an animal offering assistance to a relative is increasing the chance that the genes they have in common will survive. The greater the genetic overlap between individuals, the greater the incentive to cooperate, rather than to compete.

Hamilton's theory of kin selection was published in 1964, but its importance was not recognised by biologists at large for a further ten years. Later it was popularised by the Oxford zoologist Richard Dawkins, in his book, *The Selfish Gene*. Suddenly, social behaviour could be explained as 'selfishness in action'.

The ultimate examples of cooperative communities are found in the sea, where salps and siphonophores live. They are pretty, diaphanous creatures, sometimes shot with iridescent colours. The Portuguese man o' war trails tendrils which sting painfully should bathers touch them. Although superficially they resemble ordinary jellyfish, each is a confederacy of

A planktonic siphonophore, the ultimate example of perfect teamwork. This 'creature' is a clone, and the polyps are divided into castes specialised for particular tasks. Above, a hydroid colony

polyps working harmoniously together. Perfect team work is possible because the whole colony is a clone formed by the budding of a founder polyp. All are therefore genetically identical. This condition has enabled the polyps to become divided into classes designed for particular tasks. The sole duty of some is to feed and provide nourishment for the whole colony; others are shaped like swimming bells which pulsate and drive the structure along; some are equipped with savage stings. These kinds of polyps are infertile. However, a further caste of polyp is specially built for breeding, and these alone pass on copies of genes identical to those harboured by the barren ones. Selfless labour for the common good makes perfect sense in clones!

Insect socialism is among the greatest achievements of organic evolution. The sheer diversity and abundance of some kinds of gregarious insects beggars the imagination; driver ants in Africa may generate massive colonies of twenty-two million workers, weighing twenty kilogrammes, which patrol an area of fifty thousand square metres. Such a 'super-organism' is the creation of a single queen!

Societies of the kind seen in bees, wasps, and ants, with their caste systems and non-breeding workforce, are made possible by the peculiar method of sex determination in these insects. A queen can control the sex of her offspring by withholding sperm from her eggs. From fertilised eggs, she produces daughters – they develop into workers or future queens, depending upon the diet they receive as larvae. The queen makes sons from unfertilised eggs. This has a profound effect upon the genetic ties within the family, and between the generations. In these creatures, sisters are far more closely related to each other than they are to their own offspring, sharing seventy-five per cent and fifty per cent of each other's genes respectively. In certain circumstances, it pays a female to relinquish her ability to breed, and instead to lavish care and attention on her younger sisters, some of whom will grow into immensely fertile queens. The sterile worker therefore propagates a large proportion of its genes by proxy! Even suicidal behaviour against vicious robbers can be passed on to the next generation through sister queens, providing that the attack on the thieves was triumphant.

Hamilton's formulae and figures had a devastating inevitability about them. Given the unique genetics of bees, wasps, and ants, their colonies should evolve implacably towards states of advanced collaboration between sisters, with all the trappings of neutered, differentiated castes.

Sisterly love. A red bulldog ant worker offers an egg to a larva – one of her younger sisters. Such eggs are infertile and provide nourishment for the brood

Within the bee family (*Hymenoptera*), a full range of social life exists from primitive solitary species to the compulsively gregarious honey-making bees and carnivorous wasps. Among these colonial varieties, the balance between collaboration and competition varies from one species to another. For example, in the brown paper wasp, an open-combed species which lives in north America, each colony is founded by a single female. Shortly after she constructs a cell or two from chewed-up plant fibres, she is joined by several other females who become auxiliary queens. Besides bringing in food for the original queen, they try to introduce their own eggs into the cells. The foundress, however, eats them as soon as they are laid. After a time, the ovaries of the auxiliary queens regress, whereupon the females are employed simply as workers, cleaning the comb, fetching food, regurgitating nutrient for the grubs, and helping to build fresh cells. Recent research has raised the suspicion that these worker wasps may well be related to the founder queen, and so the cooperation is based upon kinship.

This primitive kind of wasp is not particularly aggressive, and the workers are not given to acts of bravery in defence of their queen and colony. The reason for this may be that the worker, in this species, has a remote chance of being promoted from rags to riches. Should the founding queen die, a power struggle ensues between her helpers, and the winner starts laying her own eggs, assuming the royal role. It therefore pays a female worker brown paper wasp to hedge her bets, and not to throw away her life needlessly. Among bees and wasps, the willingness of workers to engage in potentially fatal sorties against raiders seems to be related to the size of the workforce. A large colony of brown paper wasps may comprise some two hundred or so adults. Honey bees and tropical hornets possess vast squadrons of workers which sally forth at the slightest provocation, often incurring dreadful casualties in the fray. Perhaps the value of the individual is downgraded to compensate for the massive benefits of saving a few exorbitantly fertile sisters.

In the European wasp, which builds enclosed paper nests the size of cabbages, the queen keeps her daughters under control by producing chemical messages – or pheromones – which are spread through the workforce by mutual grooming. Mutual feeding also takes place between the larvae and their adult but sterile sisters; the adults provide the growing grubs with protein-rich flies and bees in exchange for sugary saliva produced by the brood, which serves as fuel for flying.

A colony of paper wasps. Like many insect colonies, it is founded upon one fertile queen and her offspring

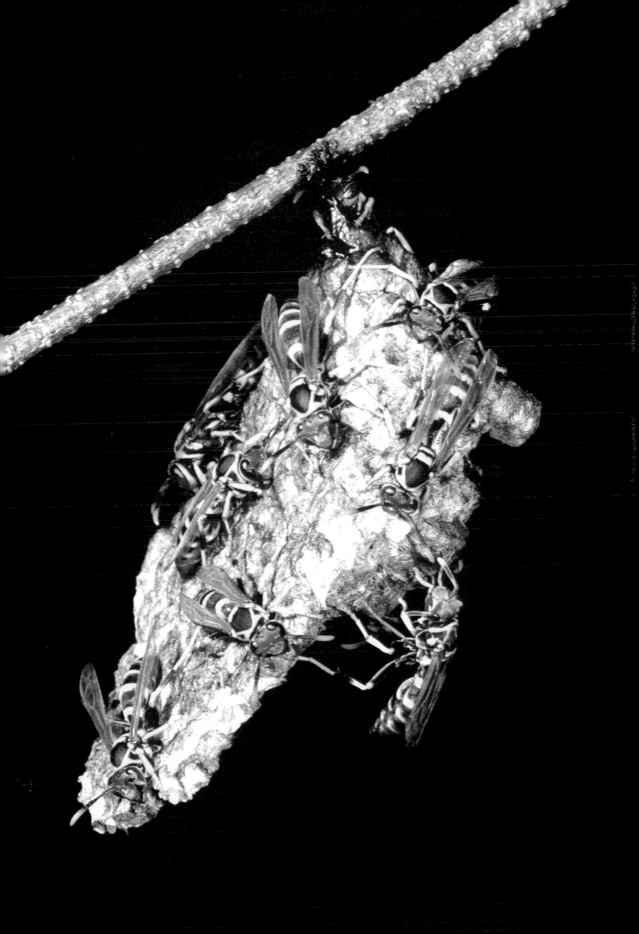

It has recently been discovered by Jennifer Jarvis, a South African zoologist, that naked mole rats live in insect-like societies, with a rudimentary caste system. These rodents would win no prizes in a zoological beauty contest. They resemble pink, animated cocktail sausages with stubby legs. As their name suggests, they have no fur, and are sightless. The most noticeable feature of their grotesque bodies is a shovel-shaped mouth with enormously prominent incisor teeth which they put to devastating use when gnashing their way through the soil, in which mole rats function superbly well. Occurring in the hot, arid regions of Ethiopia, Somalia, and Kenya, they live in extensive burrow systems which enable the creatures to tackle nutritious roots and tubers in safety, concealed from the prying eyes of predators.

The most unusual and unexpected aspect of naked mole rat behaviour is the way the labour of their communities is roughly divided between workers of different castes. By far the most industrious animals belong to what has been christened the 'frequent worker' caste. They are relatively small individuals who wriggle their way backwards and forwards along the tunnels, foraging, and transporting the food to the communal nest. They also organise themselves into gangs of 'navvies', with a leader scraping soil from the blind end of a burrow, while others kick and heave the spoil backwards and so eventually onto the surface. The 'infrequent workers' are a little larger than the real labourers, and engage in energetic tasks with markedly less enthusiasm than their smaller brothers and sisters. Members of the 'non-worker' caste are the largest mole rats in the community, spending most of their time huddled in the central nest where they may help to create a comfortable temperature for the breeding female. Besides forming the core of a living central heating system, they come into their own when the subterranean burrow system is invaded or damaged. They then spring into rare action, sounding the alarm, and are first on the scene of the disturbance. Perhaps these 'non-workers' constitute a kind of 'Praetorian guard', equivalent to the soldier caste among termites.

Both sexes participate in the mole rat caste system, but only one female breeds. She is similar in size to a 'non-worker', and devotes herself to rearing and suckling pups in the communal nest chamber. Should the nest be disturbed, the 'workers' remove the brood, restoring them to their mother when all is calm. During the weaning process, the young are brought vegetable food by the small labourers. Evidence gathered by

Jenny Jarvis suggests that, as the young mature, they enlist first as 'frequent workers'. A few remain small and never relinquish this role, whereas slightly faster-growing ones become 'non-workers'; some of these have the potential to replace the breeding female. While she is fit, the 'queen' suppresses the sexuality of her daughters, perhaps through potent perfumes released in her urine.

The mole rat appears to be a rather exceptional creature. In most higher animals, sibling sisters are no more closely related to each other than they are to their parents or to their own offspring. Competition therefore tends to flavour the relationships within groups; among birds and mammals it pays to have one's own babies, and to assist others only if there are very great benefits to the helpers in consequence.

Hamilton's theory of kinship selection had a marked effect upon those studying the social behaviour of higher animals. It focused their energies on the measurement of genetic relationships between individuals, and on looking for signs of nepotism and of acts helpful to others. With their vision sharpened by Hamilton's genetic equations, they found that nature's claw was sometimes less soaked in blood than was previously believed.

Working in the Sinai Desert, the Israeli zoologist, Amotz Zahavi, teased out the amazing intricacies of the family life of the Arabian babbler – a loose-feathered brown bird the size of a thrush, which inhabits the acacia scrub in that hot and desolate part of the world. They are highly sociable birds, which do everything together, even defending their own group territories against other parties of trespassing babblers. By assiduously catching and colour-banding babblers over a great swathe of desert, Zahavi was able to trace the relationships of the birds within his study area.

A clan of twenty or so birds was basically an extended family. On average, each contained a couple of mature cocks and a pair of hens, together with a handful of yearlings and fledglings. The senior breeding hens were sometimes sisters, and the cocks brothers, but unrelated to their mates. Both of the hens used the nest, and produced a joint clutch of up to eight eggs. All of the adults participated in incubating them, but when the chicks hatched, the non-breeding yearlings assisted their elders in supplying food for the brood. When the chicks finally left their nest, the task of looking after them fell solely upon the yearling helpers. Those fledglings which survived the rigours of desert life, graduated into helpers themselves, and supported their parents in raising further broods of brothers and sisters. Eventually they either inherit their parents' freehold, or leave to seek their own territories elsewhere. But by that time, they have already eased copies of some of their own genes into the next generation by rearing several nests of siblings.

The benefits of being a 'helper' have been accurately assessed by Patricia Moehlman, an American zoologist who studied black-backed jackals in Tanzania. She discovered that young jackals often stayed on with their parents after weaning, and assisted them to rear their next litter. Having a 'helper' appears to make all the difference between life and death for several pups. A female jackal gives birth to as many as four pups at a time, but with only her mate to support her, she rarely manages to raise more than one. However, if one of her offspring from a previous litter helps her by bringing food back to the den or by keeping hyenas away, then the chances of her pups surviving improve dramatically. In these circumstances, between two and three pups live. Like a babbler a jackal is just as closely related to its brothers and sisters as it is to its own pups. From the point of view of the genes that it carries, the young jackal is adequately rewarded for delaying its own attempts at breeding. If it 'left home' to find a mate of its own, it could only expect to raise one pup; by joining forces with its parents, its own contribution results in the rearing of an extra one or two siblings. The puzzle that Patricia Moehlman now has to answer is why young jackals bother to leave their parents at all.

Behaviour predicted by Hamilton's kinship theory is not always generous, as in the case of 'helpers'. Promoting one's own genes may be best achieved by preventing those of rivals from spreading. This can lead to habits that appear undesirable in human eyes.

Among lions the nucleus of each pride is a select circle of sisters who are joined by one or more males for several years. The males are generally brothers, and, as in the case of babblers, unrelated to the females with whom they will mate during their tenure of the pride. Sooner or later, the brotherhood will be ousted by a younger and stronger set of males.

Lion watchers in Tanzania noticed that a fresh takeover of a pride by a gang of brothers was often followed by an outbreak of cub murder by the victors. Behaviour of this kind is in the brothers' interest. They are programmed by the forces of evolution to make their own litters and not to waste time consorting with females who are suckling those fathered by their predecessors. By slaughtering the cubs, the lionesses are

Overleaf: A pride of lions on a kill. The females are probably all sisters and the males brothers, but unrelated to the lionesses

A jackal 'helper' and its younger brother from a later litter

Hanuman langurs drinking. The females often join forces to protect their young against a newly installed male

brought forward into breeding condition, and the new order of males can begin to sire their own offspring.

Hanuman langurs also practise 'infanticide', according to Sarah Hrdy, an American who has kept long-term records of the troops which live on Mount Abu, in Rajasthan, India. Langurs are elegant, nimble monkeys the size of Alsatian dogs, but sleeker and covered with silky, silver hair. From the mass of information Dr Hrdy has collected, an intriguing picture has emerged of the langur's social life, which also involves a subtle 'battle of the sexes'.

Langurs travel in parties of between twenty and thirty animals. Like gibbons, they are territorial, and early in the morning, they sit in the tree-tops whooping at each other to establish where neighbouring groups are located. Each group consists of breeding females and their infants, together with a single mature male who has the mating monopoly within the group. But he pays a price for being the troop leader, because he is beleaguered by young males from bachelor troops who aspire to take his place. He is accordingly harassed with merciless regularity, whenever his troop encounters an all-male band. If he is not on his toes, he is chased into the bush where he will spend the rest of his days in celibate isolation. For this reason, most males manage to retain their status as leaders for an average of only twenty-one months.

When a new male assumes the leadership of a troop, he is faced with a number of lactating females, heavily committed to caring for young not of his making. His strategy is to embark on a course of surreptitious infanticide, and to mate as soon as possible with his newly-acquired wives. Time is of the essence, because he can only expect at the most two years of breeding ahead of him.

But losing young to bullying males is not in the mother langur's interest. In evolutionary terms, it would make sense if they took measures to protect their genetic 'investment'. Sarah Hrdy's observations point to the fact that mother langurs have indeed developed their own tactics to reduce the risk of their infants being killed by replacement males. The presence of young monkeys is a powerful bonding force in langur troops. They are greatly admired by females. The mothers pass their young around, and only when they squirm and cry do the mothers take them back off the 'aunts'. This intense interest by the females in their offspring assures the infant monkey of a protective environment in which all of its needs are met until it becomes independent. Should a male threaten an infant, the females band together and often force the murderously inclined leader to back down. Some mothers even risk leaving the troop until their own infants are weaned.

But even motherly love has its limits. Sarah Hrdy has seen signs that there is a conflict of interest between the mother and her precious baby. Although the female langur strives to protect her offspring, she is also driven to produce as many young as possible. The sooner she can wean her baby, the sooner she can become pregnant again. However, the infant attempts to remain with its mother for as long as possible, if only to avail itself of a good supply of milk. Here lies one of the major causes of strife in langur society. As all mammals seem to do, the mother monkeys start to discourage their infants from suckling before the youngsters themselves feel inclined to forsake their mothers' breasts. When the infants are about one year old, the rejected little creatures throw temper tantrums which disrupt the prevailing air of calm which normally characterises langur troops.

Looking back over the past century, our understanding of social behaviour has changed to a remarkable extent. In the early days, zoologists scrutinised artificially constituted captive groups of animals, discovering dominance hierarchies, and individuals locked together in fear and lust. Later, researchers studied natural groups in the wild, and saw social behaviour as a benevolent activity, 'good for the species'. Nowadays, social behaviour is considered to be the outcome of a balance between conflicting self interests, with animals manipulating, deceiving, and occasionally murdering each other, in order to pass their genes on into the next generation. It is a view more in accord with Darwin's original conception of the struggle for life.

Cape hunting dogs tear at a gazelle, a part of the struggle for life whereby genes may be passed on to the next generation

THE LAST WORD?

Scientists have learned to communicate with apes by using sign language

In terms of genuine understanding of animal behaviour, are we really any wiser than Aristotle? The answer to that question must be 'yes', although we may not be as close to the ultimate truth as we think.

The boundaries of knowledge, like those of the Universe, are ever expanding. While there are still people with the patience to sit and study animals, we will continue to build upon what we know. Every day, fresh facts come to light. And yet, there are problems relating to animal behaviour which have been studied for a long time, but still remain unsolved. The mystery surrounding the navigational ability of birds fascinates us now, just as it fascinated Frederick II over seven hundred years ago. We know that birds fly by the sun and stars, are sensitive to magnetism, and can detect the pattern of polarised light in the sky. However, we remain largely ignorant about how birds apply this information to the practical purpose of finding their way from one part of the world to another. The animal mind is also just as much of a mystery to us as it was to George Romanes. Some naturalists are attempting to cull clues about the inner workings of the animal mind from the analysis of the songs and calls of advanced creatures, like dolphins and whales. Such research raises the possibility of fulfilling the dream of Dr Dolittle, of holding meaningful dialogues between animals and ourselves. The first faltering footsteps have already been taken in the USA, where several researchers have taught apes to communicate through signs and symbols. The aim of such projects is to probe the ape's potential for evolving grammatical languages, like our own.

Recently, research into animal behaviour has reflected an

The roots of artistic appreciation. When shown a series of pictures a macaque prefers Mickey Mouse!

increasing preoccupation with the necessity to understand the roots of our own behaviour. To this end, scientists internationally are using our fellow creatures as human substitutes in experiments designed to explore the learning process, the nature of aggression, the effects of overcrowding, and so on. Monkeys and apes have even been employed in the analysis of the nature of artistic appreciation. Nick Humphrey, a Cambridge zoologist, recently measured the interest a rhesus monkey took in pictures. The monkey could control the amount of time each image was projected on to a screen. He found that monkeys spent longer staring at a bold abstract painting by Mondrian than at a landscape, and it preferred pictures of unfamiliar animals to those of simple things, like bananas or daisies. Its favourite subject turned out to be a film of Mickey Mouse!

With the passing of time, well-understood behaviour is bound to be reinterpreted with the help of new ideas, spawned by the cross-fertilisation between different disciplines. This is already happening, and the relationship between ethology and experimental psychology is especially fruitful. Two decades ago, the disciples of each adopted hostile postures as they attempted to establish the boundaries of their subjects. Each argued to maintain the 'truth' of their respective points of view on the relative importance of *nature* and *nurture* in the development of behaviour. But bridgeheads have been established between the rival factions, and today scientists are much more broadminded about the necessity for seeking common ground between allied subjects.

A *détente* between ethology and ecology has likewise generated valuable insights into social behaviour. For example, the differences between closely related species cannot be understood unless the constraints of the environment and of the distribution of food are taken into account. John Crook, a British zoologist working in Africa and Asia, carefully surveyed the ninety varieties of weaver birds, some living in evergreen forests and others in open grasslands. He found that the woodland species tended to be insectivorous, highly territorial, and monogamous, contrasting with the grain-eating, highly social and polygamous savannah species. It seemed to him that the differences were shaped largely by ecological considerations. Seeds were comparatively rare in the forests, so the weavers which lived there adapted to an insectivorous diet. Since insects were distributed fairly evenly throughout the habitat, each pair of weavers had become territorial to guarantee its own food supply. Quite different conditions prevailed

on the savannahs. Grass seeds were common, and often abundant, but usually very patchy in their occurrence. A social
system based upon territorial pairs would therefore be imprac-
ticable – a territory safeguarding a fickle and patchy food
supply would need to be so vast that it would be indefensible.
The grassland weavers therefore hunted in flocks so that each
bird could capitalise on the food found by others. Shortage of
trees in such open places favoured the evolution of communal
nesting; in village weavers, the nests were huge – like haystacks – and thatched to keep the interiors cool. With nests
packed so closely together, Crook found that the aggressive
males were able to dominate several nest sites at the expense of
the less forceful cocks, so favouring a system of polygamy.

Ecological considerations have also been used to reinterpret
the lek displays of birds like black grouse and sage grouse.
There is increasing evidence that the hens which come to
copulate are lured less by the splendour of the dominant male's
rituals, and more by the quality of his property   the small
patch on which he struts. His display is more important in
warding off his rivals than in attracting a mate.

Details of this kind will continue to accumulate in scientific
journals, eventually filtering through to popular accounts of
animal habits. It is, however, difficult to predict where new
areas of research will open up, and how new information about
animals will affect our attitude towards both them and ourselves. But there are some clues buried in the story of the
discovery of animal behaviour. They show that our interpretation of what we see often reflects ideas at large in society as a
whole.

When the world was considered to be the manifestation of a
Divine Plan, it was natural for people to see animals and their
habits as parts of that scheme. The advent of evolutionary
theory tended to shift the emphasis away from religious explanations. But some scholars have suggested that the theory of
natural selection may have been affected by the climate of
laissez-faire capitalism current in the Europe of Darwin's day.
For decades after *On the Origin of Species* was published, nature
was observed as 'red in tooth and claw'. But, by the middle of
the present century, nature was being viewed more benignly.
When searching for possible reasons, it is tempting to suggest
that the rise of Behaviorism and the fascination zoologists came
to feel for communities and the apparent altruism animals
displayed was influenced by the success of Socialism as a
political force.

The current trend is to frame questions and seek answers to social problems in purely economic terms. It may therefore be no accident that contemporary accounts of animal behaviour sometimes read like economic texts, crowded with jargon and formulae. These are a far cry from the elegant descriptions which flowed from the pens of Gilbert White and Niko Tinbergen. The shift from colourful and often evocative prose is partly connected with the trend towards scientific precision, whereby actions are reduced to measurements. But it may also be due to the invasion of biology by the language and ideas of economics and mathematics, which enables scientists to look at animal behaviour from a new slant.

It is fashionable among some zoologists to apply cost-benefit analysis to behaviour. This is a mathematical technique for discovering how an animal, when faced with choices, comes to select its course of action. For instance, a blue tit must harvest an insect on average every three seconds throughout a winter's day in order to stay alive. With competing demands on its time, for, say, preening and calling, it has to make 'decisions' about what to eat, where, and for how long. The search for the tit's optimum feeding strategy can be made with the assistance of a computer. The circumstances which bear upon the bird's feeding success are reduced to an equation – a mathematical model. This will include such factors as the number of insects available, the population density of competing birds, and the distances between the trees or feeding sites; all can be assessed by good, honest field work. The computer is then enlisted to digest the figures, and from the equation, can predict how long a blue tit should feed at a particular patch, depending upon how much food is there. The scientist can then check the bird's behaviour against the prediction. If they correspond, the mathematical assumptions can then be said to model the behaviour; if not, then something which affects the tit's strategy must have been omitted, and it is a case of back to the calculus!

Sophisticated analysis of this kind brings the study of animal behaviour very firmly within the fold of science. After all, the ultimate goal of scientific endeavour is to reveal the nature of things so thoroughly that it becomes possible to forecast the way in which any system will react to a given set of conditions. Computer modelling of animal behaviour does just that, predicting not only feeding strategies, but also reproductive and fighting tactics with enormous success.

John Maynard-Smith of Sussex University has used the mathematics of 'Game Theory' to elucidate fighting strategies.

He considers fighting as a 'game', in which the contestants 'play' to win, thereby endowing more offspring than their rivals. The simplest form of game involves two animals, one a very aggressive individual which fights viciously, a 'hawk', and the other a timid 'dove'. Maynard-Smith reckons that the way an animal plays the game depends upon what its opponent does. In a population of animals which tend to settle their disputes only by bluff and harmless ritual, a 'hawk' which fights fiercely, perhaps killing its rivals, will quickly dominate the scene. Winning all of its fights, it will breed successfully, and produce many more murderous male offspring. But eventually, 'hawks' become so common that now a less favourable situation exists for them. A 'hawk' will be faced with others of similar disposition, and so stand an even chance of being killed or badly injured in a quarrel. In this situation, the balance tips in favour of those with 'dove' strategies. By quickly turning tail and fleeing at the first sign of a fight, they are more likely to retain their life, and to breed. Clearly, neither a population of 'hawks' or 'doves' is stable. Both can be 'invaded' by animals adopting opposite strategies. Maynard-Smith is able to calculate what mixture of 'hawks' and 'doves' constitutes an evolutionary stable strategy (ESS) for any given species, depending upon its way of life. The ramifications of Game Theory are very complex, but, in essence, the concept of ESS has been enormously useful in helping us to understand how disputes can be settled by force or ritual.

Modern methods of expressing problems encountered in animal behaviour do not imply that animals themselves are calculating tacticians. They are not. If their behaviour lends itself to this mathematical kind of analysis, it means that natural selection has bestowed upon them precise behaviour patterns for dealing with perfectly predictable environmental situations.

In the past, naturalists simply watched animals and marvelled at their behaviour. Over the centuries, people became increasingly curious, and used their observations to formulate theories to explain that behaviour. Today's zoologist builds upon the knowledge of his distinguished predecessors, constructs his mathematical models, and only then looks to the animals to confirm or deny the correctness of his assumptions. Perhaps this technique will develop and eventually will generate quite different kinds of stories to those presented in this book.

Whichever way we look at animals, they are a source of unending delight, and will continue to beguile generations of scientists and naturalists to come. A great deal has already been achieved in understanding our fellow creatures, and, perhaps through them, we have come to know ourselves a little better. However, as our impression of the world will undoubtedly change, we will continue to apply new techniques and fresh ideas to the understanding of the animal kingdom.

FURTHER READING

The following books deal with various aspects of animal behaviour in a detailed and systematic way.

ALLEN, D. ELLISTON (1976). *The Naturalist in Britain*. Allen Lane, London. An interesting social history of the development of nature study.

BASTOCK, M. (1967). *Courtship: a zoological approach*. Heinemann, London. A good account of this, perhaps the most intriguing category of behaviour.

BOAKES, R. (in press). *From Darwin to Behaviourism*. Cambridge University Press, Cambridge. A fascinating survey, with psychological leanings, of the origins of the science of animal behaviour. Easy to read.

DAWKINS, R. (1976). *The Selfish Gene*. Oxford University Press, Oxford. A best-selling, popular account of Hamilton's ideas of 'kin selection'.

GOODALL, J. (1970). *In the Shadow of Man*. Collins, London. The story of Jane Goodall's relationship with the chimpanzees of Gombe.

HINDE, R. A. (1970). *Animal Behaviour: a synthesis of ethology and comparative psychology*. (2nd ed.) McGraw-Hill, New York. A bible to all students of animal behaviour, although it is heavy going for the general reader.

HINDE, R. A. (1982). *Ethology: its nature and relations with other sciences*. Oxford University Press, Oxford. Basically an undergraduate text, but less formidable than *Animal Behaviour*. Hinde is one of the most important of modern scientists and theoreticians, and anything he writes is of the utmost significance.

JENKINS, A. C. (1978). *The Naturalists. Pioneers of Natural History*. Hamish Hamilton, London. A well-illustrated survey, including accounts of such people as John Ray, Charles Waterton and many others.

KREBS, J. R. and DAVIES, N. B. (1978). *Behavioural Ecology*. Blackwell, Oxford. Although a book for the expert, this does present a very modern view of behaviour, in terms of 'economic strategies'.

LORENZ, K. (1960). *King Soloman's Ring*. (2nd ed.) Methuen, London. Autobiographical account of Lorenz's early years at Altenburg. It gives a wonderful glimpse of his rapport with animals.

LORENZ, K. (1979). *The Year of the Greylag Goose*. Eyre-Methuen, London. Lavishly illustrated with photographs taken by Lorenz's goose watchers.

MANNING, A. (1979). *An Introduction to Animal Behaviour*. Arnold, London. Although primarily written for university students, Professor Manning's text is lucid and takes a broad biological approach. He deals, among other things, with the genetics and evolution of behaviour.

MCFARLAND, D. (1981). *The Oxford Companion to Animal Behaviour*. Oxford University Press, Oxford. This is an important book for all those interested in animals. Seventy or so distinguished experts have contributed essays on a range of subjects, such as 'sleep', 'communication', 'hormones' and the 'brain'. The history of the study of animal behaviour is also dealt with in detail. Although a work of reference, it is a mine of information and eminently readable.

FURTHER READING

RACHLIN, H. (1976). *Introduction to Modern Behaviourism*. (2nd ed.) Freeman, Reading. A good introductory text to the subject.

SCHALLER, G. (1963). *The Mountain Gorilla*. University of Chicago Press, Chicago. The work that rocked the zoological world of the 1960s, when Schaller showed that wild mountain gorillas could be studied at arm's length.

TINBERGEN, N. (1953). *The Herring Gull's World*. Collins, London. A New Naturalist Monograph.

TINBERGEN, N. (1972). *The Animal and its World*. (2 vols.) Allen & Unwin, London. A chance to read Tinbergen's most important scientific papers, including reports of his hunting wasp, stickleback and gull experiments.

TINBERGEN, N. (1972). *Curious Naturalists*. Penguin Books, London. A revised edition of an earlier book. A delightful read, part autobiography and part a popular rendition of scientific work. It includes a chapter on the cliff-dwelling kittiwakes, and Esther Cullen's research.

VON FRISCH, K. (1954). *The Dancing Bees*. Methuen, London. The fascinating story of von Frisch's work with bees, dealing with their language, their sensitivity to scents and to ultra-violet and polarised light.

WILSON, E. O. (1971). *The Insect Societies*. Harvard University Press, Harvard, Mass. Probably Wilson's best book, a classic, lucid and well-illustrated study of the weird and wonderful world of insect communities.

WILSON, E. O. (1975). *Sociobiology: the new Synthesis*. Harvard University Press, Harvard, Mass. A hefty but attractive book, surveying social behaviour throughout the animal kingdom. It has caused an uproar among sociologists who do not adhere to the view that social behaviour is the outcome of genes.

Various biographies are also available of such people as Ray, White, Darwin, von Frisch, Watson, Skinner and Lorenz. Many books pertinent to the discovery of animal behaviour are long out of print, but nevertheless remain classics. The following may turn up in junk shops or in catalogues of antiquarian books.

DARWIN, CHARLES (1872). *The Expression of the Emotions in Man and Animals*. John Murray, London.

FRASER DARLING, F. (1937). *A Herd of Red Deer*. Oxford University Press, Oxford.

MORGAN, L. H. (1970). *The American Beaver and His Works*. (Facsimile ed.) Burt Franklin, New York.

ROMANES, G. (1882). *Animal Intelligence*. Kegan Paul, London.

TINBERGEN, N. (1951). *A Study of Instinct*. Oxford University Press, Oxford.

WHITE, GILBERT (1789). *The Natural History and Antiquities of Selborne*.

ZUCKERMAN, S. (1932, reprinted 1981). *The Social Life of Monkeys and Apes*. Routledge & Kegan Paul, London.

ACKNOWLEDGEMENTS

This book has been written during the production period of a television series on the same theme. In consequence, all of those who applied their skills so generously to the series contributed directly or indirectly to the book. The scale of the teamwork involved is conveyed on the next page, where their names are recorded in gratitude.

My thoughts on the subject matter of this book have been greatly enriched by many people, both within and outside the BBC. Foremost among them is my colleague, Michael Salisbury, who had the task of taking three of my scripts and bringing them to life. This he did so delightfully and sensitively, that they surpassed my ambitions for them. I am also grateful to cameraman Jim Saunders who translated our directions into superbly evocative drama sequences. Natural history programmes largely stand or fall on the quality of their wildlife pictures. Many specialised natural history cameramen from all over the world provided film for *The Discovery of Animal Behaviour*. I should like to acknowledge my debt to these dedicated people who, at my behest, spent weeks and sometimes months waiting for the right moments to squeeze the triggers of their cameras. Some of their 'shots' appear in this book.

This project demanded a vast amount of research. I greatly benefited from conversations with naturalists, zoologists, and experimental psychologists who so willingly shared their views and memories with me. It is therefore a pleasure for me to record my sincere thanks to: Prof. G. P. Baerends, Dr P. Bateson, Dr R. Boakes, Dr John Bowlby, F.R.S., Drs M. & E. Cullen, A. Fraser Darling, Prof. Karl von Frisch, P. Gautrey, Dr J. Gould, Prof. W. Hamilton, Prof. H. Harlow, Dr K. Heinroth, Prof. R. Hinde, F.R.S., Dr J. van Iersel, Prof. K. Immelmen, Dr J. Krebs, Prof. M. Lindauer, Prof. K. Lorenz, Dr S. Lovari, Prof. P. Marler, Prof. D. McFarland, Prof. N. McKintosh, Dr L. Rogers, Dr P. Sevenster, Prof. B. F. Skinner, Prof. S. Suomi, Prof. N. Tinbergen, F.R.S., Dr S. Tomas, Prof. W. H. Thorpe, F.R.S., Prof. E. O. Wilson, Lord Zuckerman, F.R.S. I hope I have heeded their counsel and, where relevant, presented their research wisely within the brief of the series.

Marion Zunz played a crucial part in the preparation of this book. Not only did she provide much of the background information upon which the stories were constructed, she also read the initial drafts of the manuscript. Her advice was invaluable. I likewise valued the assistance of Pelham Aldrich-Blake, whose experience of studying monkeys and apes in tropical forests enhanced both the television series, and the book, especially Chapter 7. I should also like to record my thanks to Robert MacDonald of Collins, and Toby Roxburgh, of BBC Publications, for their helpful criticism of the manuscript; the text is all the better for their advice; to Jennifer Fry for finding such attractive illustrations, and to Peter Campbell, who designed this book.

Finally, I am indebted to Sara Clements, whose unflagging spirit was a source of encouragement when, as from time to time is bound to happen, it seems more tempting to do nothing than to wield the pen or tap out a lively tune on the typewriter.

J.H.S.

THE DISCOVERY OF
ANIMAL BEHAVIOUR

SERIES PRODUCER	John Sparks
PRODUCER	Michael Salisbury
ASSISTANT PRODUCERS	Andrew Buchanan, Pelham Aldrich-Blake
RESEARCH	Marion Zunz
PRODUCTION ASSISTANTS	Jane Trethowan, Ruth Levy
NARRATOR	Andrew Sachs
CAMERAMEN	Jim Saunders, Richard Ganniclifft, David Haylock, Wolfgang Bayer
WILDLIFE CAMERAMEN	Hugh Miles, Ron Eastman, Rodger Jackman, Martin Saunders, Hugh Maynard, Walter Deas, Bob Brown, Jim Frazier, Theo Cockerel, Gerald Thompson, Owen Newman, Peter Scoones, Maurice Tibbles, Chris Mylne
ASSISTANT CAMERAMAN	Red Denner
FILM LIGHTING	Ron Robinson, John Palmer
FILM EDITOR	Ron Martin
DUBBING EDITOR	Alec Brown
SOUND	Roger Long, Lyndon Bird, Andy Nelson
MUSIC	Edward Williams
DESIGNER	Sally Hulke
PROPERTY BUYERS	Richard Geuter, Sally Clements, Robin Rumbelow
GRAPHICS DESIGNER	Peter Netley
MAKE-UP ARTISTS	Lucy Hutchinson, Daphne Barker
COSTUME DESIGNER	Catriona Tomalin

Picture credits

Picture research: Jennifer Fry

INDEX